Computer-Managed Maintenance Systems

Computer-Managed Maintenance Systems

Editor
_____

Manoj Karkare

scitus
academics

Computer-Managed Maintenance Systems

Edited by **Manoj Karkare**

Printed in 2017

ISBN: 978-1-68117-348-1

Library of Congress Control Number: 2015939261

© 2016 by

SCITUS Academics LLC,
616, Corporate Way, Suite 2, 4766,
Valley Cottage, NY 10989

www.scitusacademics.com

Printed in United States of America on Acid Free Paper ∞

Contents

Preface

Effective resource management and reliable equipment are essential for optimum plant performance. Computer-Managed Maintenance Systems goes beyond the simple selection and implementation of a CMMS. It also defines the changes in infrastructure, management philosophy and employee skills that must be implemented to gain maximum benefits from the CMMS. The book is designed to address the information needs of all levels of plant management.

Editor

Aluminium Process Fault Detection and Diagnosis

Nazatul Aini Abd Majid[1] Mark P. Taylor[2] John J. J. Chen[2], and Brent R. Young[2]

[1]Center for Artificial Intelligence Technology, Universiti Kebangsaan Malaysia (UKM), 43600 Selangor, Malaysia

[2]Department of Chemical and Materials Engineering, The University of Auckland, Auckland 1142, New Zealand

ABSTRACT

The challenges in developing a fault detection and diagnosis system for industrial applications are not inconsiderable, particularly complex materials processing operations such as aluminium smelting. However, the organizing into groups of the various fault detection and diagnostic systems of the aluminium smelting process can assist in the identification of the key elements of an effective monitoring system. This paper reviews aluminium process fault detection and diagnosis systems and proposes a taxonomy that includes four key elements: knowledge, techniques, usage frequency, and results presentation. Each element is explained together with examples of existing systems. A fault detection and diagnosis system developed based on the proposed taxonomy is demonstrated using aluminium smelting data. A potential new strategy for improving fault diagnosis is discussed based on the ability of the new technology, augmented reality, to augment operators' view of an industrial plant, so that it permits a situation-oriented action in real working environments.

INTRODUCTION

A variety of fault detection systems for the aluminium smelting process can be found in the literature. This diversity is contributed to principally by the way in which each system utilizes the resources available by using an approach which is appropriate for the process control system in question. Investigating these systems by identifying elements that shape the systems may help us to understand the different kinds of fault detection system in the aluminium smelting process. Thus, we classified these elements in the following groups.

- Fault detection and diagnostic knowledge: what knowledge is used in the fault detection and diagnosis systems of the aluminium smelting process?
- Fault detection and diagnostic techniques: how is the system built by utilizing the knowledge?
- Usage frequency: how frequently can the system monitor the process?
- Results presentation: how are the results of the system presented to the operators?

The aim of this work is to identify taxonomy of aluminium process fault detection and diagnosis system with four key elements: techniques, knowledge, usage frequency, and results presentation. This work also aims to identify the potential ability of augmented reality as one of the techniques in results presentation.

This paper will first describe a fault detection and diagnostic taxonomy that has been developed from reviews of the literature and knowledge pertaining to the aluminium smelting process. Secondly, the groups and elements that comprise the taxonomy are explained. Next, the key elements of the new system for the aluminium smelting process that have been identified in this research based on the taxonomy are discussed and demonstrated with an example. Finally, in order to further assist in fault diagnosis, the integration of augmented reality that can be used as a potential new strategy is discussed.

THE PROPOSED TAXONOMY FOR ALUMINIUM PROCESS FAULT DETECTION AND DIAGNOSIS

The groups and elements that create a fault detection and diagnostic taxonomy for the aluminium smelting process are illustrated in Figure 1. The proposed taxonomy can assist in determining the various factors in developing a new fault detection and diagnosis system. The groups and elements of this taxonomy are briefly described in the following section.

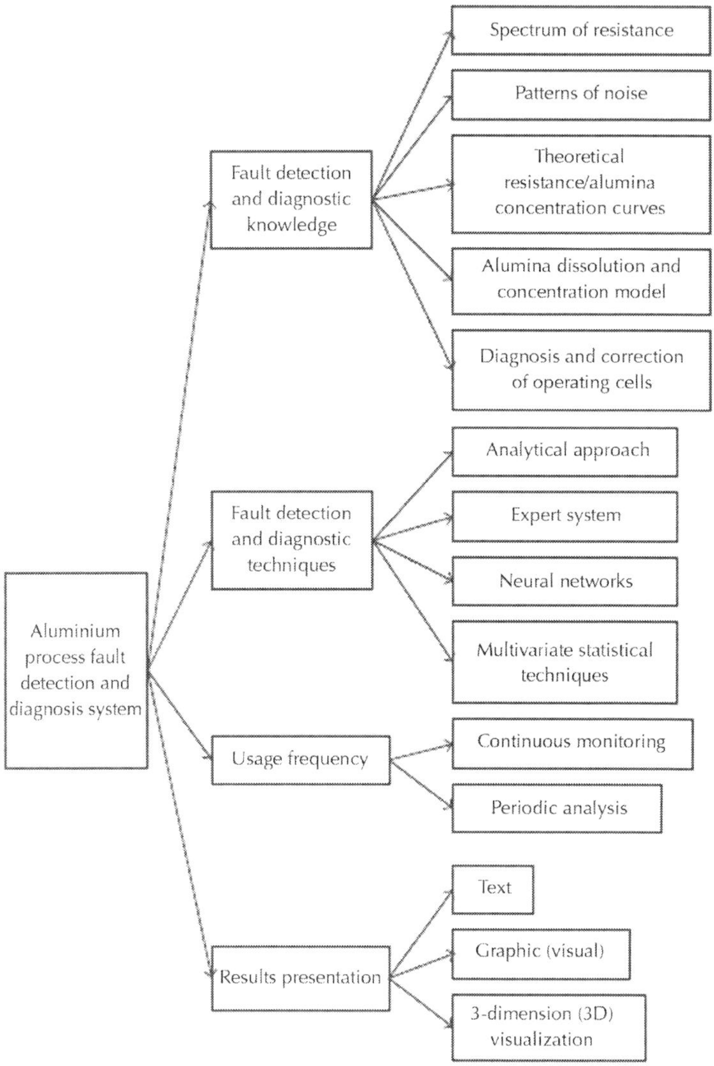

Figure 1: Taxonomy for aluminium process fault detection and diagnosis.

Fault Detection and Diagnostic Knowledge

The first group is comprised of fault detection and diagnostic knowledge. The elements of this group represent particular knowledge in the aluminium smelting process that has been used, and can be used, to develop fault detection and diagnosis systems. A brief explanation of each element is given below.

- The first element in this group is a spectrum of resistance in which the specifications of the spectra in three cases were identified for assisting in fault diagnosis. The cases are normal cell, aluminium roll, and abnormal anode [1].

- Patterns of noise constitute the second element in this group. Three different patterns of noise were recognized, to assist in fault diagnosis. These are bubble noise [2], short-circuiting noise, and metal pad roll noise [2, 3].

- The third element in the group is a theoretical resistance/alumina concentration curve. There have been researchers who have selected data for developing fault detection systems by using this curve as an important reference such as Meghlaoui et al. [4], Yurkov et al. [5], and Nagem et al. [6].

  The first example stems from research by Meghlaoui et al. [4] in which two dynamic trend indicators were generated based on the theoretical resistance/alumina concentration curve.

  The second example comes from research carried out by Yurkov et al. [5] in which selected data were deemed appropriate for analysis based on feeding cycles. These cycles were formed following the controlling of alumina feeding based on the theoretical resistance/alumina concentration curve.

  The third example is from research by Nagem et al. [6] in which data were divided into four regions based on the theoretical resistance/alumina concentration curve. These regions were (1) lean (low alumina concentration), (2) normal (good operating point), (3) rich (high alumina bulk concentration), and (4) very rich (high alumina bulk concentration, high temperature, and reoxidation phenomena). These examples from previous research indicate the difficulties experienced in the direct measurement of alumina concentration and the frequent measurement of important parameters, such as cell temperature; this has prompted a group of researchers to discover how to utilize existing knowledge in the development of an appropriate fault detection and diagnosis system.

- The fourth element is a set of colour and textural features grouped according to the varying alumina content of anode cover materials. These colour and textual features were identified using multivariate image analysis techniques. These features can be used to estimate the alumina content of anode cover materials [8].

- The fifth element is the diagnosis and correction of operating cells that were recorded by operators and engineers. This knowledge can be used to form a knowledge database in an expert system (e.g., [9, 10]). It can also assist in the discovery of new knowledge for fault diagnosis and then for validating that new knowledge by using the procedure for knowledge discovery from databases (e.g., [11]).

Fault Detection and Diagnostic Techniques

The development of fault detection and diagnosis systems involves not only various knowledge domains but also a variety of methods. In the taxonomy proposed here, the group pertaining to the techniques to be used for fault detection and diagnosis is described as the second group. This group concerns the development of a fault detection system by using a suitable technique and utilizing specific knowledge. A brief explanation of each element is given below.

- The first element in this group is an analytical approach because two common methods for this approach, parameter estimation and diagnostic observers, were used to develop an aluminium process detection system [12]. The approach was based on a quantitative model in a well-accepted taxonomy developed by Venkatasubramanian et al. [13] in which precise first principles or mathematical models of the process are used to model a system based on the relationship between the inputs and outputs of the process. The differences between actual system behaviour and that of the system model are then calculated and called residuals [13].

- Figure 2 shows the two main stages in model-based fault detection and diagnosis [7] where some of the frequently used residual generation methods are diagnostic observers, Kalman filters, and parameter estimation. These residuals are further evaluated in order to identify the occurrence of faults in the process [7].

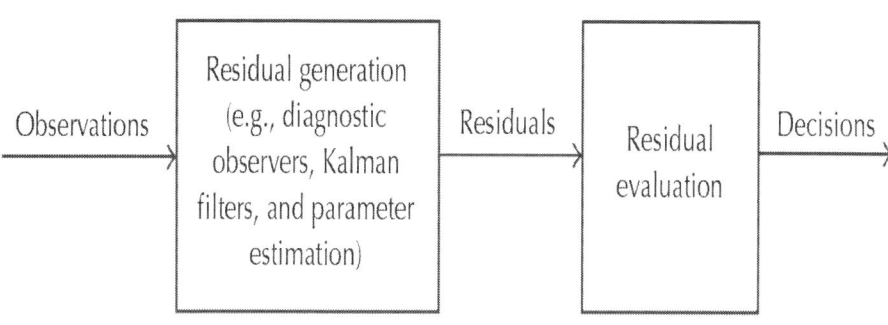

Figure 2: The two main stages in model-based fault detection and diagnosis (redrawn from [7]).

In a fault detection system for the aluminium smelting process, an extended Kalman filter was used in order to not only estimate the alumina concentration in different sections of an aluminium reduction cell, but to also indicate an abnormal alumina distribution. A mathematical model was developed to estimate the alumina concentration. Residuals were generated from the difference between the alumina concentration expected by the system model and the actual concentration [12]. Abnormal alumina distribution was detected when the residuals were significant. However, the residuals not only indicate abnormal events but may also indicate other sources including noise, disturbances, and model errors [7]. This issue of robustness may limit the effectiveness of using the Kalman filter or other model-based approaches.

- An expert system which is a process history-based approach is the second element in this group. In the process history-based approach, prior knowledge is extracted from a large amount of historical data. This feature extraction can be divided into qualitative and quantitative methods as shown in Figure 3 [14]. A popular example of a qualitative method is the expert system where prior knowledge from experts is extracted to represent human knowledge in a particular domain. It is used in fault diagnosis to infer a conclusion of an out-of-control situation by combining the facts from a user with the knowledge from human experts represented in knowledge databases. In the aluminium smelting process, knowledge relating to diagnosis and correction of operating cells was incorporated in a number of expert systems such as those of Haldris [9], the FMFA-based expert system [10], and the CVG Venalum potline supervisory system [15]. In an aluminium electrolysis process expert system (AEPES) [16], for example, there were two subsystems; the first one incorporated more general knowledge of the aluminium reduction cell including unstable cell voltage, anode carbon quality, and higher iron impurity. The second one incorporated specific knowledge including bath temperature, metal level, and bath ratio. The use of an expert system, however, lacks statistical inference and pattern recognition [17].

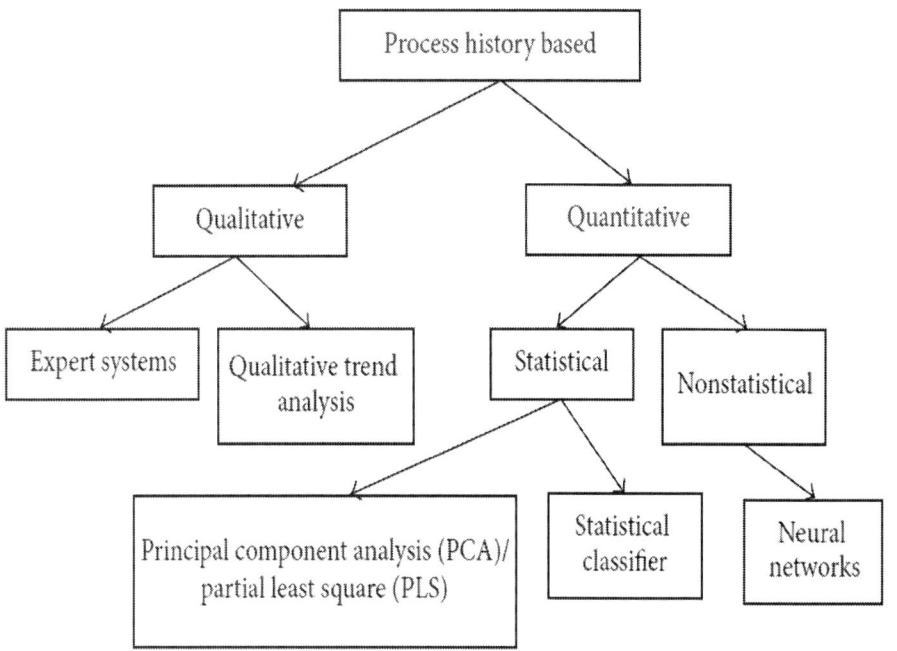

Figure 3: Classification of process history-based methods (redrawn from [14]).

The third element in this group, neural networks, is also a process history-based approach. As shown in Figure 3, the quantitative method can be divided into statistical and nonstatistical. The use of artificial neural networks is a nonstatistical approach used in fault diagnosis to recognise the received pattern of data by using a nonlinear mapping between input (data patterns) and output (fault classes). This mapping consists of hidden neurons that are highly interconnected and arranged in layers [17]. In the aluminium smelting process, a backpropagation neural network was used to map spectra of cell resistance and output vectors for three cases which were normal cell, aluminium roll, and abnormal anode [1]. In addition, a feedforward neural network was used to predict cell resistance and as a fast dynamic indicator [4]. Both systems used simulation data to train the networks. The use of neural networks, however, lacks the ability to generalise/explain behaviour [18].

- The fourth element in this group is the use of multivariate statistical techniques which is also a quantitative and process history-based approach. Multivariate statistical techniques such as PCA and PLS are used to extract a number of latent variables from normal operating data which are retrieved from historical databases, in order to form an empirical model [19, 20]. Thus, in the future, whenever the behaviour of the operation of the plant differs from the empirical model of the normal process, unexpected changes in the process can be detected [20]. The following are examples of the use of PCA/PLS for process monitoring in materials processing including aluminium processing:

  a combination of PCA and linear discriminant analysis (LDA) was used for monitoring the quality of iron and steel [21];

  PCA was used for monitoring the quality of copper [22];

  multivariate image analysis was used for estimating alumina concentration on anode cover [8];

  estimation was carried out for aluminium reduction cell performance using PLS [23];

  (e)multivariate monitoring of aluminium reduction cells was undertaken using PCA [24];

  multivariate online monitoring of preheating, start-up, and early operation of aluminium reduction cells was investigated using PLS regression [25].These examples show that the multivariate techniques, PCA/PLS, have been investigated for analysis of historical data and monitoring of processes in various complex process industries because of their ability to handle large volumes of highly correlated data.

Usage Frequency

The third group to be considered in this taxonomy is usage frequency where it applies to the way the fault detection and diagnosis system performs its analysis of the process. A brief explanation of each element is given below.

- The first element in this group is one which is continuous. An online fault detection and diagnosis system monitors the process continuously by analysing continuous data from the process. The system may immediately signal abnormal events after they happen. Examples of these systems include a backpropagation neural network developed by Shuiping et al. [1] and a feedforward neural network system developed by Meghlaoui et al. [4].

- Periodic analysis is the second element in this group. In an aluminium smelting process, an offline fault detection and diagnosis system periodically analyzes data at a frequency ranging from daily to once in two days. This level of frequency is to enable the detection of abnormal events using bath chemistry and heat balance parameters. Some of the examples in this system include (1) process monitoring using PCA [24] and (2) an analytical model for estimating alumina concentration and abnormal events [12].

Results Presentation

The fourth group in this taxonomy describes the three modes for presenting the detection results: text, graphics (visual), and three-dimensional (3D) visualization. The presentation of detection results to the operator can be more informative if the operator's needs are considered in terms of a clear visual indication in the screen design [26]. This theory is supported by research done by Harris et al. [27] where colour and statistical graphs were incorporated in the design of a supervisory control system. The use of the bold, contrasting colour in this system clearly indicates when there is an alarm so that the section leader in a smelter can act accordingly. Furthermore, the potential operator colour sensitivity should be considered by choosing colour palettes that provide effective contrast for all potential colour vision levels.

Although many contributors to the reference literature pertaining to aluminium process fault detection cited in this taxonomy do not provide screen design in their approach, there are some articles that do provide or describe how the results are presented. Three major examples are given here. Firstly, in a multivariate statistical application, a 3D visualization was used to illustrate Hotelling's T^2 statistic with a 3D control envelope which is based on bath temperature, liquidus point, and cumulative sum of alumina feed ratios [28]. Secondly, a fault diagnosis system based on a neural network had a screen interface in which two modes of presentation were used: text (querying history report, spectrum analysis of cell resistance, and fault diagnosis for the cell state) and graphics (real-time curve and history curve of the cell signals) [1]. Thirdly, a supervision system for aluminium reduction cells based on mathematical models had an interface displaying five functions including real-time display and curve change for specified parameters [29]. The state of the cells is displayed using a text box, and the temperature, the voltage, the current, and the alumina concentration are displayed in charts. The user interface also consists of control boxes, such as combo boxes, and control buttons. These three examples show that a combination of text and graphics may be more effective for revealing monitoring results to the operator than solely using either text or graphics [29].

A fault detection and diagnosis system will now be shown and discussed in the next sections in order to demonstrate how a new system can be developed based on the proposed taxonomy.

CASCADE FAULT DETECTION AND DIAGNOSIS SYSTEM

The cascade fault detection and diagnosis system [30, 31] was designed to detect any faults and then diagnose faults that are related to anode effect, anode spike, block feeder, and low alumina dissolution. This system is presented as an example of how faults such as an anode effect can be detected and diagnosed with multivariate statistical techniques as can be seen in Figure 4. The key elements of this system are discussed below by referring to Figure 4.

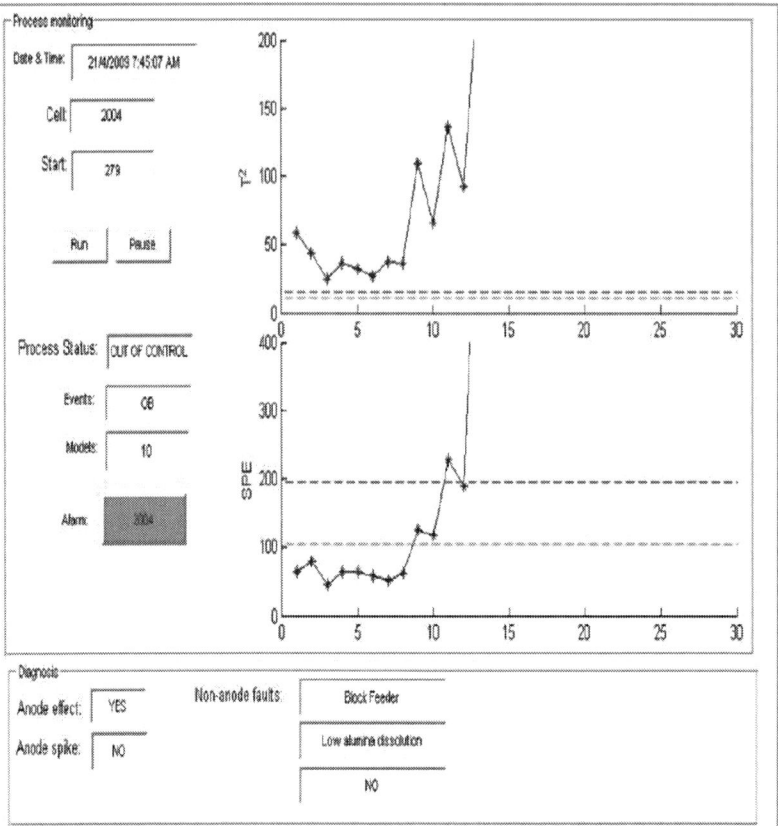

Figure 4: Operator screen shows an indication of an anode effect and its possible causes: a block feeder and low alumina dissolution.

Fault Detection and Diagnostic Knowledge

The first element of this system is the discovery of new knowledge based on the established relationship between pseudoresistance and alumina concentration. In addition to the extraction of knowledge from the prior research of experts and the production of a theoretical resistance/alumina concentration curve through experiment, learning to identify abnormal patterns from data is one of the practical ways by which to discover fresh knowledge relating to fault detection and diagnosis [32]. Since there is a need to develop a fault detection and diagnosis system based on the changes of cell voltage and cell resistance patterns within overfeed/underfeed cycles, ascertaining abnormal patterns within these cycles using data mining to discover new knowledge was carried out in this research [30]. In Figure 4, for example, abnormal patterns within the current cycles were detected and diagnosed to be related to the patterns of an anode effect and patterns of the previous cycles to be related to a block feeder and low alumina dissolution.

Fault Detection and Diagnostic (FDD) Techniques

In the first element of the system, the established relationship between pseudoresistance and alumina concentration is used as a basis for discovering new knowledge. In the second element, the established relationship is used as the basis for monitoring the process with the added use of a suitable FDD technique. It has been interesting to note that the established relationship between pseudoresistance and alumina concentration has become the basis for many applications from linear to nonlinear models for a range of purposes such as (1) the estimation of alumina concentration using the Kalman filter approach [12], (2) the prediction of anode effects using a linear time-series model and a simple nonlinear exponential rise curve [33], and (3) the prediction of feed control decision variables using neural networks [4]. The strengths and weaknesses of some of these applications were discussed by Stevens Mcfadden et al. [34] where an application using the neural network model has been suggested as a suitable approach for a predictive modelling task.

As discussed above, many fault detection techniques have been employed previously. The main interest of this research, however, is a technique that is capable of early fault detection in the industrial application of the aluminium smelting process. All of the previously mentioned applications in this research stem from analytical and knowledge-based approaches, the focus of which has mainly been on the avoidance of anode effects. Less attention has been given to the use of data-driven approaches such as PCA and PLS for observing the changes of patterns within the overfeed-underfeed cycle for the detection and diagnosis of problems. Also, many researchers have used only simulated data instead of real data. Since the aluminium smelting process is complex, having many problems that require effective process monitoring, it may be impractical to develop an accurate and explicit mathematical model of the process for this purpose. Therefore, the model-based methods, both quantitative and qualitative, have not been considered in this work.

On the other hand, there has been growing interest in using process history-based approaches for fault detection of industrial applications [14, 19]. Venkatasubramanian et al. [14] listed three key reasons for this increasing interest; these are as follows: (1) they are easy to put into practice, (2) little modelling effort is required, and (3) little prior knowledge is needed. A number of process history-based fault diagnostic techniques have been developed for the aluminium smelting process including expert systems, neural networks, and multivariate statistical techniques (PCA/PLS). Firstly, a number of expert systems were developed for fault diagnosis [9, 10]. Due to the complexity of the aluminium smelting process, the cause of an abnormal operating pattern is often difficult to diagnose. Process engineers may interpret the abnormal pattern themselves before or while using an expert system. A computerized system that is capable of solving the persistent problem of diagnosing abnormal patterns for multiple aluminium reduction cells is needed. Furthermore, expert systems require considerable effort in order to build a knowledge-based diagnosis system for a complex and large process. An existing solution for this problem has been based on neural networks [1]. However, this requires comprehensive and excessive amounts of data, causing Shuiping et al. [1] to use simulation data instead of real data in their study. The use of PCA and PLS is a viable option because only moderate amounts of historical data are needed. Based on this, an application of PCA was developed by Tessier et al. [24] for monitoring the aluminium electrolysis process. However, in order for a monitoring system to be rendered effective, consideration needs to be given to dynamic cell behaviour.

Therefore, the objective of this system is to incorporate the dynamic behaviour of the two important events of anode changing and alumina feeding during the aluminium smelting process, for effective and timely fault detection and diagnosis. This can be done by using a new multivariate statistical framework using PCA and PLS. In this system, PCA has been chosen for the development of a fault detection system and a combination of PCA and PLS has been chosen for the development of a system for fault diagnosis. This is because these multivariate statistical techniques can address some of the problems arising in the detection and diagnosis of faults in the aluminium smelting process.

- Firstly, PCA or PLS can handle a substantial quantity of data which is both correlated and noisy.

- Secondly, both PCA and PLS use a noncausal model so that the lack of a causal model in the aluminium smelting process is not an issue. A causal model needs a first principles model.

- Thirdly, multiway PCA (MPCA) and multiway PLS (MPLS), extensions of PCA and PLS, respectively, are able to handle any nonlinear behaviour during the process of alumina feeding.

- Finally, PCA and PLS are effective in practice for the monitoring of the aluminium smelting process since the reference models have been mainly built from process data [35].Principally, the use of multivariate statistical techniques such as PCA and PLS needs to be investigated not only for the prediction of anode effects, but also for the diagnosis of problems that cause anode effects and for the early detection of anode spikes. This advanced monitoring of aluminium processing leads to a reduction in energy consumption and emission of PFCs. Abnormal patterns within the alumina feeding cycles were analysed using MPCA and MPLS. As shown in Figure 4, the monitoring charts used in the system were based on MPCA. These charts are Hotelling's T^2 chart and the SPE chart. The abnormal events detected by these charts were then diagnosed using MPLS in order to classify patterns related to these abnormal events [31].

Usage Frequency

The continuous monitoring of changes of variability patterns within the overfeed/underfeed cycles is preferred in this research for early fault detection and diagnosis [30]. As shown in Figure 4, five-minute data were used for monitoring the process. The monitoring charts detected and diagnosed an anode effect 25 minutes before it occurred in the real operation. This shows an early detection and diagnosis of an anode effect.

Results Presentation

Charts that can show changes of pattern against acceptable limits for operations are one of the important elements in monitoring.

Information about the current process and the results of the diagnosis that were provided in textual form were put together with the charts. In this research [30, 31], a mixture of text and graphics incorporated with suitable colour (red and green) and user control boxes such as a combo box for selecting cells was used instead of selecting only one mode in order to demonstrate clearly abnormal events. In Figure 4, for example, the operator's screen indicated this situation by a change in the colour of button for cell 2004 from green to red, the status of the process from "IN CONTROL" to "OUT OF CONTROL," and the status of the anode effect detection from "NO" to "YES." A clear indication of abnormal events as shown in this example can help process engineers and operators to timely respond to problems that occur in the process.

The Need of Augmented Reality (AR)

Augmented reality is a viable option for improving the results presentation. Results from the system were mostly based on computer-generated information such as text, graphics, charts, and tables. Operators in the smelters will take actions based on this information. Integrating this digital information with a real situation might help further in fault diagnosis. In fact, this is the basis of augmented reality where it has been defined in a broad sense as augmenting natural feedback to the operator with simulated cues [36]. The main reason for using AR is its capability of augmenting a user's view of an industrial plant, so that it permits a situation-oriented action in real working environments. The integration of augmented reality within an aluminium process fault detection and diagnosis system is a potential new strategy for improving decision making in fault diagnosis.

THE NEW STRATEGY FOR ALUMINIUM PROCESS FAULT DETECTION AND DIAGNOSIS

A new strategy for fault detection and diagnosis was proposed to incorporate AR technique. This adds a new element in terms of results presentation as shown in Figure 5. AR was selected because it is a novel human-computer interaction tool that overlays computer-generated information on a real-world environment [37]. This technique has been applied in industry, for example, the Boeing wire harnessing project [38], car engine maintenance [39], and an intelligent welding gun [40]. These works have shown potential of AR to be combined with human abilities to offer efficient and complementary tools to assist manufacturing tasks [37]. This motivates this research to propose a new strategy for improving fault detection and fault diagnosis in the aluminium smelter.

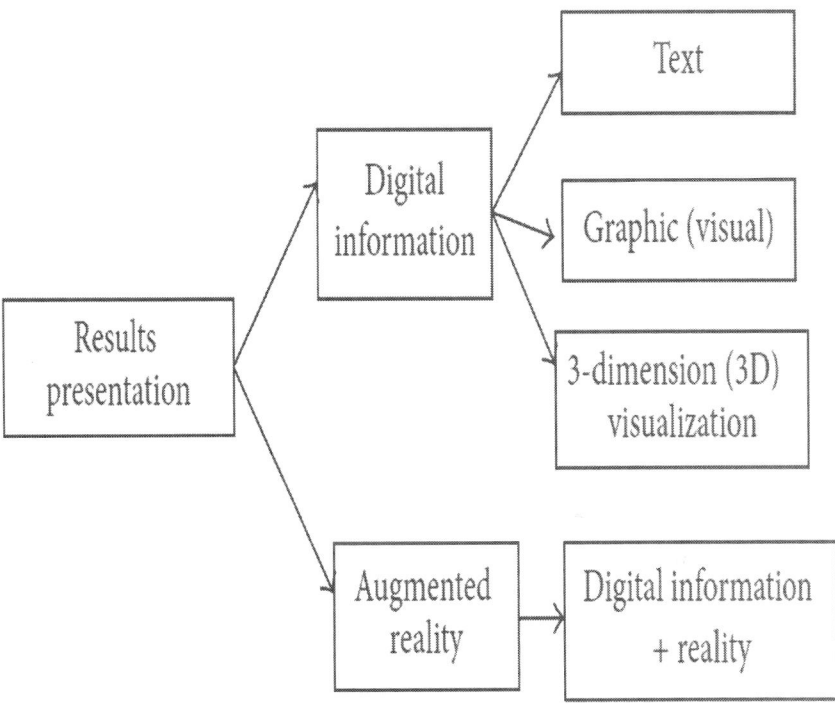

Figure 5: The addition of new elements, augmented reality, in the detection mode of results.

Procedure for the New Strategy

In this new strategy, there are five steps in incorporating AR in the fault detection and diagnosis system: requirement, design, development, implementation, and evaluation. These steps are described in the sections that follow.

Requirement

In the first step, the requirements for AR technology for a specific task in manufacturing are identified. This identification is based on the need for error-free job execution, reduced cognitive load, ease of learning a task [37], and assisted decision making. When the specific task has been identified, the current industry situation needs to be studied in order to support the task, in terms of its end-users, level of expertise, and current environment [41]. One of the tasks of operators in an advanced supervisory control and management system (named integrated potline control and improvement, referred to as IPC-Im hereafter) is root causes diagnosis [42]. This task can be combined with AR (as illustrated in Figure 6) in order to offer efficiency of information presentation and to assist developer of the system in providing an improved interaction between human and the system.

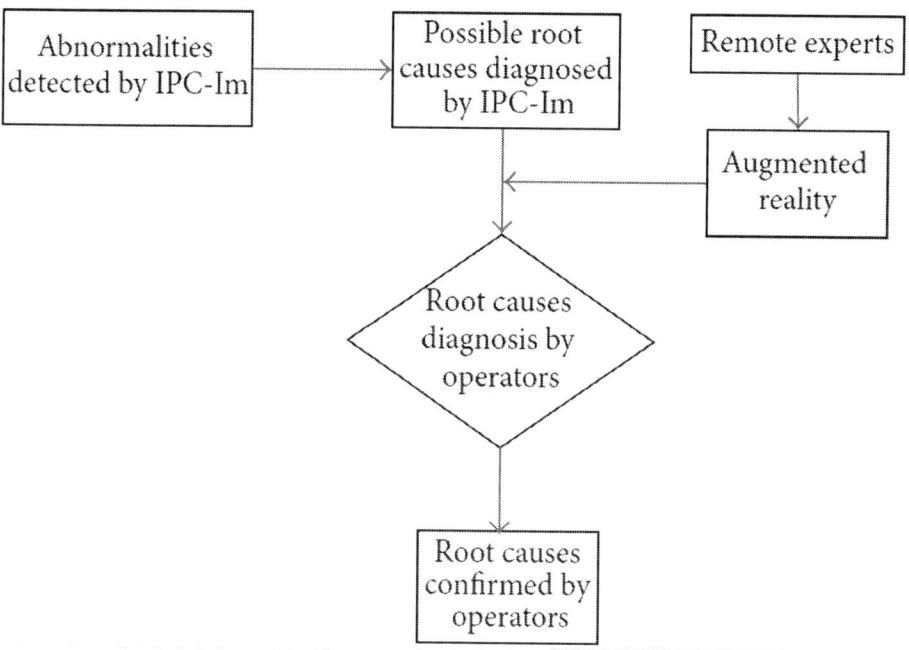

Figure 6: Detection tool with AR application (adapted from [42]).

Design

A number of the main elements, which were identified from AR-assisted maintenance system by Nee et al. [37], can be considered in this design phase. These elements are as follows.

Display Device. What device is used for visual output? Examples include head-mounted display (HMD) (e.g., [43, 44]), handheld devices (HHD) (e.g., [45, 46]), and projectors (e.g., [47]).

Since handheld devices, such as mobile phones, can be used as a tool to view this information overlay, mobile AR has gained increased attention from academia and industry, due to the portability of mobile phones and the ubiquitous nature of camera phones [37]. Therefore, mobile phone is one of the viable options as a display device in this new strategy.

Tracking Technologies. What technology is used for tracking the cameras position, in order to register virtual objects? Examples include vision-based tracking (marker) (e.g., [48]), sensor-based tracking (e.g., [49]), or a hybrid (i.e., vision- and sensor-based tracking) (e.g., [50]). In this new strategy, vision-based tracking (marker) has been used to simulate how information of a cell can be superimposed on a live view of the cell. An example of such markers is shown in Figure 7, where the camera first locates the marker (which in this case is an image of aluminium reduction cell's number, 2053). When the marker is recognized, a superimposed image (shown as information of a cell, e.g., temperature, excess AlF_3, liquidus temperature, and voltage) will appear on the screen, in order to mix the virtual world with real world that is being viewed. This innovative technology offers a solution for assisting in monitoring a complex process industry, such as the aluminium smelting industry.

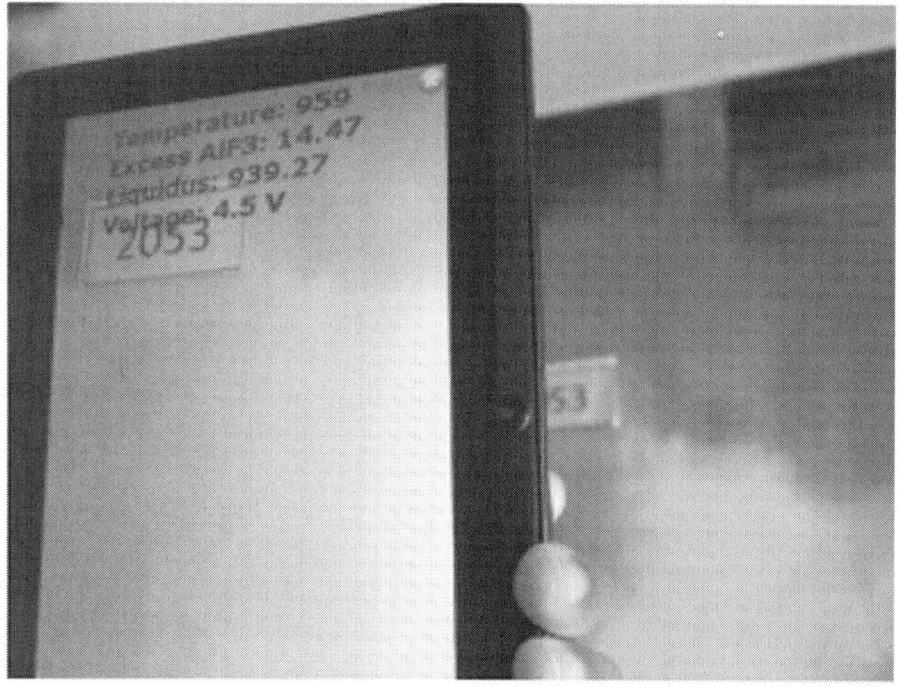

Figure 7: Augmented view of cell 2053 with superimposed information for four process variables.

Data Management. How data is managed in the AR-based application? Examples include scenario-oriented/process-oriented (e.g., [51]), knowledge-based (e.g., [50]), or virtual model-based (e.g., [52]) data retrieval.

Human-System Interaction. How do humans interact with the AR-based system? Examples include using mouse, keyboard, microphone, touchscreen (e.g., [53]), and digital camera.

User Collaboration. How does an AR-based application provide collaboration among users? Examples include using a microphone and a remote laser pointer (e.g., [52]).

If abnormalities can be diagnosed using the proposed mobile AR-based approach, a real-time fault diagnosis system could be developed as an advanced tool to diagnose problems in an aluminium smelter as shown in Figure 8. In this application, all the processing work and file saving can be done in the cloud of the Internet after considering the limited processing capability of the mobile phone [37].

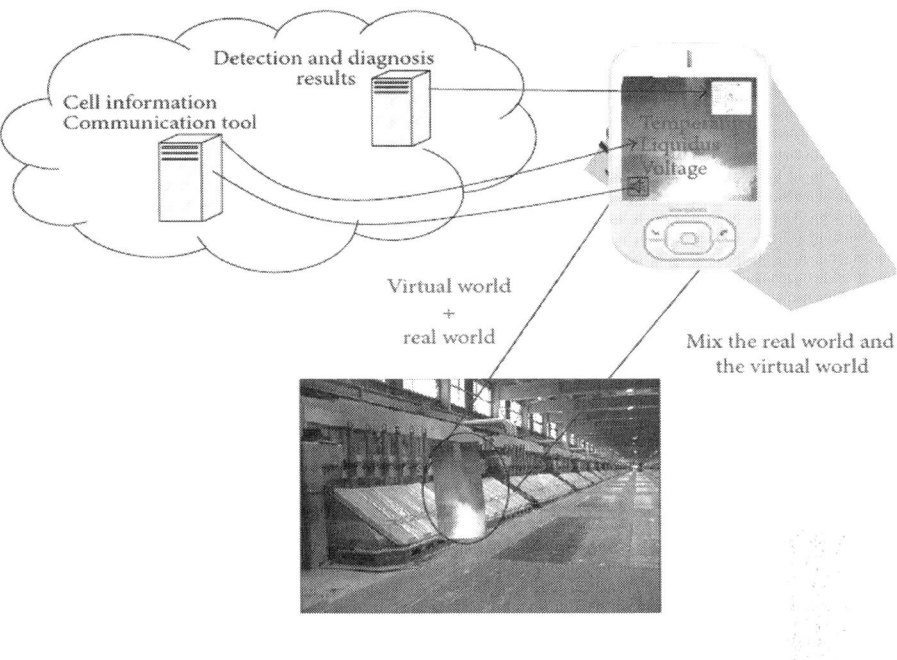

Figure 8: Abnormalities potentially generate a distinct view within an aluminium reduction cell. The combination of this view and current information of the cell can be used as a guideline to diagnose problems.

The mobile AR module should provide sufficient information for the process operators to diagnose operating problems. Six functions that need to be considered are the plant information system, linking of documents, machine history, interactive troubleshooting, error tracking and feedback, interactive video, and a virtual laser pointer. The potential view of the mobile AR module for a process operator in an aluminium smelting plant is illustrated in Figure 9 where there are four main functions:

- buttons for interactive manipulation,
- speech-based interaction,
- results from diagnosis module,
- cell information, current status, and faults' history.

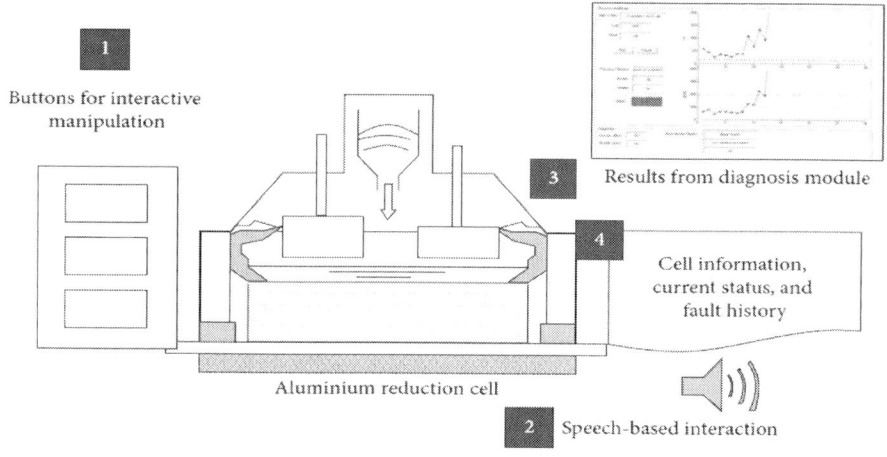

Figure 9: Augmented view of the real world with superimposed information for four main functions.

Development, Implementation, and Evaluation

The third step is development where the platform used to build the AR application is selected as being either a browser or a device platform. This development should focus on the usability and performance of the application. In the fourth step, implementation, possible problems in setting up the application should first be identified [41], before implementing the application. A clear action plan should be developed in order to assist end-users or workers in using the new technology. The fifth step is evaluation where user satisfaction is evaluated, and the benefits of the application are identified.

These five steps (requirement, design, development, implementation, and evaluation) can be used as guidance in developing an AR application for any manufacturing plant, such as the aluminium smelting plant. In addition, the AR module can also be added in corrective action guidelines because AR can be used to highlight dangerous area in a plant. A virtual fire in the plant, for example, might help an operator to have in-depth understanding with operating procedures when an abnormal situation occurs. Therefore, operator behaviour in normal and abnormal situation can be tested in order to improve operating procedures.

CONCLUSIONS

Developing a fault detection and diagnosis system for the aluminium smelting process is a major challenge. This fault detection and diagnosis system should be able to accurately indicate abnormal situations although the process is complex and dynamic. In this paper, the proposed taxonomy described with examples of existing systems was given. The taxonomy clearly highlights the key elements of a fault detection and diagnosis system which covers utilization of knowledge, FDD techniques, usage frequency, and results presentation. The taxonomy has many uses including the following:

- to identify the key elements to distinguish between existing systems,
- to identify areas of improvement for the existing systems,
- to provide an overview of the system where various techniques have been applied to detect and diagnose faults. This taxonomy has helped in the development of this work by identifying the gap in existing fault detection and diagnosis systems and realizing a new approach to developing a new system that is practical, provides timely detection and diagnosis, and is easy to understand by operators. In the future, the use of AR technology can enhance the competence of the diagnostic module to diagnose problems in a more practical manner. AR can provide an interactive environment, where operators and remote experts can communicate using the same field of vision. Since AR can be used to augment a user's view of an industrial plant, it provides alternative solutions for design, quality control, monitoring and control, service, and maintenance in complex process industries, such as the aluminium smelting industry.

REFERENCES

1. Z. Shuiping, L. Jinhong, and D. Lei, "Fault diagnosis system for 350kA pre-baked aluminium reduction cell based on BP neural network," TMS—Light Metals, pp. 583–587, 2007.

2. L. Banta, C. Dai, and P. Biedler, "Noise classification in the aluminum reduction process," TMS—Light Metals, pp. 431–435, 2003.

3. R. Bijun, W. Zhao, S. Dai, and S. Chen, "Research of fuzzy control for alumina in Henan HongKong Longquan Aluminium CO.LTD., China," TMS—Light Metals, pp. 439–442, 2007.

4. A. Meghlaoui, J. Thibault, R. T. Bui, L. Tikasz, and R. Santerre, "Neural networks for the identification of the aluminium electrolysis process," Computers & Chemical Engineering, vol. 22, no. 10, pp. 1419–1428, 1998.

5. V. Yurkov, V. Mann, K. Nikandrov, and O. Trebukh, "Development of aluminium reduction process supervisory control system," TMS—Light Metals, pp. 263–267, 2004.

6. N. F. Nagem, J. V. Da Fonseca Neto, and C. A. Braga, "Pattern identification for feed control strategy using fuzzy neural algorithm," in Proceedings of the 11th International Conference on Computer Modelling and Simulation (UKSIM ‹09), pp. 380–385, March 2009.

7. J. J. Gertler, Fault Detection and Diagnosis in Engineering Systems, Marcel Dekker, New York, NY, USA, 1998.

8. J. Tessier, C. Duchesne, C. Gauthier, and G. Dufour, "Estimation of alumina content of anode cover materials using multivariate image analysis techniques," Chemical Engineering Science, vol. 63, no. 5, pp. 1370–1380, 2008.

9. W. K. Rolland, A. Steinsnes, A. S. Larsen, and K. A. Paulsen, "Haldris—an expert system for process control and supervision of aluminium smelters," TMS—Light Metals, pp. 437–443, 1991.

10. A. I. Berezin, P. V. Polyakov, O. O. Rodnov, V. L. Yasinski, and P. D. Stont, "FMFA-based expert system for electrolysis diagnosis," TMS—Light Metals, pp. 429–434, 2005.

11. X. Z. Wang, Data Mining and Knowledge Discovery for Process Monitoring and Control, Springer, London, UK, 1999.

12. K. Hestetun and M. Hovd, "Detection of abnormal alumina feed rate in aluminium electrolysis cells using state and parameter estimation," Computer Aided Chemical Engineering, vol. 21, pp. 1557–1562, 2006.

13. V. Venkatasubramanian, R. Rengaswamy, K. Yin, and S. N. Kavuri, "A review of process fault detection and diagnosis part I: quantitative model-based methods," Computers & Chemical Engineering, vol. 27, no. 3, pp. 293–311, 2003.

14. V. Venkatasubramanian, R. Rengaswamy, S. N. Kavuri, and K. Yin, "A review of process fault detection and diagnosis part III: process history based methods," Computers & Chemical Engineering, vol. 27, no. 3, pp. 327–346, 2003.

15. C. Abaffy, R. Aiquel, J. Larez, and J. Gonzalez, "CVG Venalum potline supervisory system," TMS—Light Metals, pp. 301–305, 2006.

16. S. P. Lu, Control and supervision of the aluminium electrolysis process with expert system [Ph.D. thesis], Quebec University, 2002.

17. V. Uraikul, C. W. Chan, and P. Tontiwachwuthikul, "Artificial intelligence for monitoring and supervisory control of process systems," Engineering Applications of Artificial Intelligence, vol. 20, no. 2, pp. 115–131, 2007.

18. R. Andrews, J. Diederich, and A. B. Tickle, "Survey and critique of techniques for extracting rules from trained artificial neural networks," Knowledge-Based Systems, vol. 8, no. 6, pp. 373–389, 1995. ·

19. T. Kourti, "Application of latent variable methods to process control and multivariate statistical process control in industry," International Journal of Adaptive Control and Signal Processing, vol. 19, no. 4, pp. 213–246, 2005.

20. J. F. MacGregor, H. Yu, S. G. Muñoz, and J. Flores-Cerrillo, "Data-based latent variable methods for process analysis, monitoring and control," Computers & Chemical Engineering, vol. 29, no. 6, pp. 1217–1223, 2005.

21. M. Kano and Y. Nakagawa, "Recent developments and industrial applications of data-based process monitoring and process control," Computer Aided Chemical Engineering, vol. 21, pp. 57–62, 2006. ·

22. C. WikströmA, C. Albanoa, L. Erikssona, et al., "Multivariate process and quality monitoring applied to an electrolysis process: part I. Process supervision with multivariate control charts," Chemometrics and Intelligent Laboratory Systems, vol. 42, pp. 221–231, 1998.

23. J. Tessier, C. Duchesne, G. P. Tarcy, C. Gauthier, and G. Dufour, "Analysis of a potroom performance drift, from a multivariate point of view," TMS—Light Metals, pp. 319–324, 2008.

24. J. Tessier, T. G. Zwirz, G. P. Tarcy, and R. A. Manzini, "Multivariate statistical process monitoring of reduction cells," TMS—Light Metals, pp. 305–310, 2009.

25. J. Tessier, C. Duchesne, G. P. Tarcy, C. Gauthier, and G. Dufour, "Increasing potlife of Hall-Héroult reduction cells through multivariate on-line monitoring of preheating, start-up, and early operation,"Metallurgical and Materials Transactions B, vol. 41, no. 3, pp. 612–624, 2010.

26. I. Miletic, S. Quinn, M. Dudzic, V. Vaculik, and M. Champagne, "An industrial perspective on implementing on-line applications of multivariate statistics," Journal of Process Control, vol. 14, no. 8, pp. 821–836, 2004.

27. D. Harris, Y. Gao, M. Taylor, J. Chen, and M. Hautus, "Operational decision making in aluminium smelters," in Engineering Psychology and Cognitive Ergonomics, Springer, Berlin, Germany, 2009.

28. M. A. Stam, M. P. Taylor, J. J. J. Chen, A. Mulder, and R. Rodrigo, "Common behaviour and abnormalities in aluminium reduction cells," TMS—Light Metals, pp. 589–593, 2008.

29. Z. Shuiping and Z. Qiuping, "A supervision system for aluminium reduction cell," TMS—Light Metals, pp. 463–468, 2003.

30. N. A. A. Majid, M. P. Taylor, J. J. J. Chen, M. A. Stam, A. Mulder, and B. R. Young, "Aluminium process fault detection by Multiway Principal Component Analysis," Control Engineering Practice, vol. 19, no. 4, pp. 367–379, 2011.

31. N. A. A. Majid, M. P. Taylor, J. J. J. J. Chen, W. Yu, and B. R. Young, "Diagnosing faults in aluminium processing by using multivariate statistical approaches," Journal of Materials Science, vol. 47, no. 3, pp. 1268–1279, 2012.

32. M. P. Taylor and J. J. J. Chen, "Advances in process control for aluminium smelters," Materials and Manufacturing Processes, vol. 22, no. 7-8, pp. 947–957, 2007.

33. M. Vajta and L. Tikasz, "Adaptive prediction of anode effects in aluminium reduction cells," inSelected Papers from the IFAC Symposium, pp. 311–315, Pergamon, Oxford, UK, 1987.

34. F. Stevens Mcfadden, G. P. Bearne, P. C. Austin, and B. J. Welch, "Application of advanced process control to aluminium reduction cells—a review," TMS—Light Metals, pp. 1233–1242, 2001.

35. L. H. Chiang, E. L. Russell, and R. D. Braatz, Fault Detection and Diagnosis in Industrial Systems, Springer, London, UK, 2001.

36. P. Milgram, H. Takemura, A. Utsumi, and F. Kishino, "Augmented reality: a class of displays on the reality-virtuality continuum," in Telemanipulator and Telepresence Technologies, vol. 2351 ofProceedings of SPIE, Boston, Mass, USA, October 1994.

37. A. Y. C. Nee, S. K. Ong, G. Chryssolouris, and D. Mourtzis, "Augmented reality applications in design and manufacturing," CIRP Annals—Manufacturing Technology, vol. 61, no. 2, pp. 657–679, 2012. ·

38. D. Mizell, "Boeing's wire bundle assembly project," in Fundamentals of Wearable Computers and Augmented Reality, W. Barfield and T. Caudell, Eds., Lawrence Erlbaum Associates, New York, NY, USA, 2001.

39. H. Regenbrecht, G. Baratoff, and W. Wilke, "Augmented reality projects in the automotive and aerospace industries," IEEE Computer Graphics and Applications, vol. 25, no. 6, pp. 48–56, 2005.

40. F. Echtler, F. Sturm, K. Kindermann et al., "The intelligent welding gun: augmented reality for experimental vehicle construction," in Virtual and Augmented Reality Applications in Manufacturing, S. K. Ong and A. Y. C. Nee, Eds., Springer, 2004.

41. N. Navab, "Developing killer apps for industrial augmented reality," IEEE Computer Graphics and Applications, vol. 24, no. 3, pp. 16–20, 2004.

42. G. Yashuang, M. Taylor, J. J. J. Chen, P. Lavoie, and M. Hautus, "Advanced supervisory control of smelters," in Proceedings of the 10th Australasian Aluminium Smelting Technology Conference, 2011.

43. D. Cheng, Y. Wang, H. Hua, and M. M. Talha, "Design of an optical see-through head-mounted display with a low f-number and large field of view using a freeform prism," Applied Optics, vol. 48, no. 14, pp. 2655–2668, 2009.

44. Z. Zheng, X. Liu, H. Li, and L. Xu, "Design and fabrication of an off-axis see-through head-mounted display with an x-y polynomial surface," Applied Optics, vol. 49, no. 19, pp. 3661–3668, 2010.

45. M. Hakkarainen, C. Woodward, and M. Billinghurst, "Augmented

assembly using a mobile phone," in Proceedings of the 7th IEEE International Symposium on Mixed and Augmented Reality (ISMAR ‹08), pp. 167–168, Cambridge, UK, September 2008. ·

46. B. Stutzman, D. Nilsen, T. Broderick, and J. Neubert, "MARTI: mobile augmented reality tool for industry," in Proceedings of the WRI World Congress on Computer Science and Information Engineering (CSIE ‹09), pp. 425–429, Los Angeles, Calif, USA, April 2009.

47. M. Kitagawa and T. Yamamoto, "3D puzzle guidance in augmented reality environment using a 3D desk surface projection," in Proceedings of the IEEE Symposium on 3D User Interface (3DUI ‹11), pp. 133–134, Singapore, March 2011.

48. B. Schwald, J. Figue, E. Chauvineau, et al., "STARMATE: using augmented reality technology for computer guided maintenance of complex mechanical elements," in Proceedings of the eBusiness and eWork Conference, pp. 17–19, Venice, Italy, October 2001.

49. S. Feiner, B. Macintyre, and D. Seligmann, "Knowledge-based augmented reality," Communicationsof the ACM, vol. 36, pp. 53–62, 1993.

50. S. Henderson and S. Feiner, "Opportunistic tangible user interfaces for augmented reality," IEEE Transactions on Visualization and Computer Graphics, vol. 16, no. 1, pp. 4–16, 2010.

51. P. Savioja, P. Jarvinen, T. Karhela, P. Siltanen, and C. Woodward, "Developing a mobile service-based augmented reality tool for modern maintenance work," in Proceedings of the 2nd International Conference on Virtual Reality, pp. 554–563, Beijing, China, 2007.

52. P. Harmo, A. Halme, P. Virekoski, M. Halinen, and H. Pitkanen, "Etala-virtual reality assisted telepresence system for remote maintenance," in Proceedings of the 1st IFAC conference on Mechatronic Systems, Darmstadt, Germany, 2000.

53. J. Didier and D. Roussel, "Augmented reality assistance in train maintenance tasks," in Proceedings of the Workshop on Industrial Augmented Reality (ISMAR ‹05), 2005.

Chapter 2

Performance Assessment of Product Service System from System Architecture Perspectives

John P. T. Mo

School of Aerospace, Mechanical and Manufacturing Engineering, RMIT University, Bundoora, Melbourne, VIC 3083, Australia

ABSTRACT

New business models in complex engineering products have favoured the integration of acquisition and sustainment phases in capability development. The product service system (PSS) concept enables manufacturers of complex engineering products to incorporate support services into the product's manufacturing and sustainment lifecycle. However, the PSS design has imposed significant risks to the manufacturer not only in the manufacture of the product itself, but also in the provision of support services over long period of time at a predetermined price. This paper analysed three case studies using case study research design approach and mapped the service elements of the case studies to the generic complex engineering product service system (CEPSS) model. By establishing the concept of capability distribution for a PSS enterprise, the capability of the CEPSS can be overlaid on the performance-based reward scheme so that decision makers evaluate options related to the business opportunities presented to them.

INTRODUCTION

Recent trend around the world among the owners of complex engineering systems such as aircraft or oil refinery is to include consideration for the sustainment of the system at the very early stages of system development. According to the Defence Materiel Organisation in Australia [1], the asset acquisition project is considered a continuum of four phases, which can be generalised as a capability systems lifecycle as shown in Figure 1. The goal is to ultimately attain desired capability levels that can be measured as a performance outcome of systems in-service.

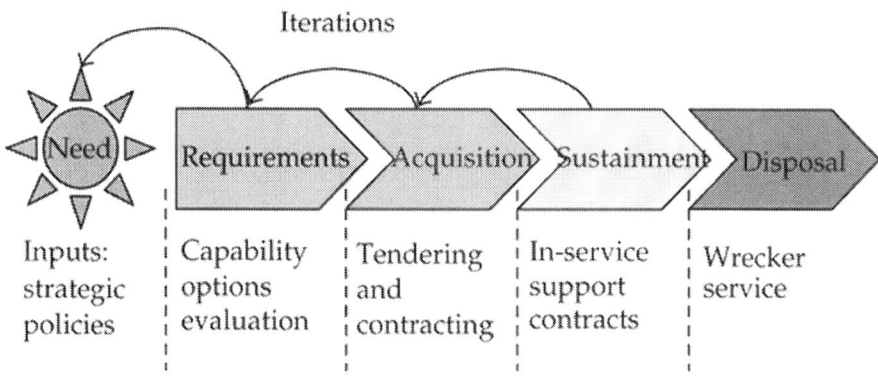

Figure 1: Capability systems lifecycle.

There are two different contracting regimes in Figure 1.

- System acquisition agreements including functional and performance specification of the final system, that is, the tendering and contracting activities in the acquisition phase.
- Sustainment agreements specifying outcomes and performance requirements for in-service support, that is, the in-service support contracts in the sustainment phase.Although it is still early stage process development, the Australian Defence intends to adopt a more integrated approach by contracting for acquisition

and sustainment simultaneously in some of their new system acquisitions.

Similarly, the Ministry of Defence [2] in UK is managing a general shift in defence acquisition away from the traditional pattern of designing and manufacturing successive generations of platforms. Instead, a new paradigm centred on support, sustainability, and the incremental enhancement of existing capabilities from technology insertions has evolved with the emphasis increasingly on through life capability management. The new approach to acquisition is built around the objective of achieving.

a. primacy of through life considerations;

b. coherence of defence spend across research;

c. development, procurement, and support,

d. successful management of acquisition at the departmental level. From the industry's point of view, this shift in defence acquisition process means longer, more assured revenue streams based on long-term support and ongoing development instead of a series of big "must win" procurements [3]. Observation on other industry sectors shows similar changes in the manufacturing and sustainment of complex engineering products such as an oil refinery and mining machinery [4].

Traditionally, management of sustainment services is the responsibility of the asset owner, after the product is commissioned. Most asset owners simply take the recommended schedule of the manufacturer, either by in house service department or by a maintenance services contractor. The strategy is to minimise expenditure that should be spent on the asset [5]. Therefore, classical services and maintenance plans are designed on the principle that mean time between failures is a constant and hence the focus is to replace components before it fails. Typically, service activities including inspection, adjustment, and replacement are scheduled in fixed intervals [6]. Due to multifaceted relationship between operating context and characteristics inherent in the complex system, these intervals may not be optimised [7]. On the other hand, reliability-centred maintenance regime has been developed to plan actions in maintenance based on advanced understanding of the reliability of the system [8]. In addition, many other factors are also influencing the operations of the asset [9]. Many service decisions on assets are therefore made on rules of thumbs rather than using analysed system performance data. Many complex systems are left vulnerable with high risks of failure.

A new service business model known as "performance-based contracting" has emerged in recent years as one of the favourable choice of contracting mechanisms for the public sector and asset intensive industries [10]. Under performance-based contracting approach, a contractor offering systems support services needs to design an operation and support system that is sustainable, fits for the purpose, and demonstrates its value for money. The advantage of performance-based contracting is the sharing of benefits for both sides of the business. Efficiency gains are shared between the contractor and the owner of the business [11]. In this regards, original equipment manufacturers have significant advantage over other service providers because they know their product well. Many equipment suppliers have taken the opportunity to expand into offering after sales services to customers [12].

A manufacturer of complex engineering system entering into this kind of contract takes a lot of risks with the contract. For example, the new contracting framework by Defence Materiel Organisation [13] contains several elements of incentives and penalties. Application of these elements depends on the actual system performance results within four bands.

- Performance Band I. System performance result is below the required performance level. Contract payment is reduced proportionally to the actual performance outcomes as a disincentive.

- Performance Band II. Result is poor but may be tolerable for short term only. Contract payment is significantly reduced proportionally to the actual performance outcomes with a more rapid reduction ratio until reaching zero.

- Performance Band III. Result is totally unsatisfactory and represents an irrecoverable failure. No payment is made and other remedies may be applied.

- Performance Band IV. Result of the achieved performance equals or exceeds the required performance level. An optional performance incentive may be paid in addition to the agreed value.

In this new service business model, risks exist throughout the whole of life of the product. How can the manufacturer know in advance what performance he/she can achieve at the conceptual design phase of the product life cycle? Which performance band is the system going to operate? There are many factors affecting system performance, for example, new deployment requirements, or change of software operating system, and so forth. The technology in the system is already very sophisticated. The model adds further complexities of sustainment and lifelong services in the commercial contract. These represent several layers of uncertainties in the commitment on the part of the service provider. A decision support methodology that can reduce such uncertainties in the life of the asset is required. This paper is motivated by the fact that new business paradigm emerging in the service sector has demanded a new set of principles and knowledge to assist the manufacturing industry. A new product and service enterprise architecture is proposed in this paper and verified by three past service systems design projects, with case study research design methodology. Based on the new architectural model, an assessment methodology that can be used for assisting decision makers to evaluate options available to them in regards to the business opportunities presented to them is proposed.

PRODUCT SERVICE SYSTEM

The shift of complex engineering products manufacture to service-oriented business environment has necessitated research in developing a new business model [14]. Abe [15] studied a service-oriented solution framework designed for Internet banking and described the new research as "service science". The concept of product service system (PSS) was initially developed around the optimization of sustainability criteria to operations, maintenance, and environmental related issues around the product [16]. The PSS concept extends, on the basis of an existing complex product, the provision of support services on that complex product when it is in operation. Bairnes et al. [17] presented a clinical style survey of contemporary practices in PSS and subsequently defined PSS as a special case of servitization. It is obvious that there are commercial benefits for companies to move into continuous services and support operations of the complex products they manufacture. In addition to the technical requirements of supporting a complex engineering system, a key feature of the new PSS model is the extension of product offering to service offering. A service system comprises people and technologies that adaptively adjust a system's value of knowledge while the system changes in its lifecycle [18].

One of the key questions emerged from this approach is the uniqueness of service requirements. Every complex engineering product is different and hence it is fair to say that each PSS is customised. Johansson and Olhager [19] examined the linkage between goods manufacturing and service operations and developed a framework for process choices that enable joint manufacturing and after-sale services operations. Study showed that moving into services-oriented business could have significant financial implications to the company [20]. In a performance-oriented service system, decisions for optimization can be quite different from maintenance oriented service concepts. For example, in order to reduce time of service to customers, Shen and Daskin [21] suggested that a relatively small incremental inventory cost would be necessary to achieve significant service improvements. Hence, to develop service systems that can handle this type of business requirements, companies should build common business functionalities as shared services so that they can be reused across lines of business as well as delivery channels [22].

When compared to traditional support arrangements, PSS concept changes a contractor's roles and responsibilities by shifting the support service to customer focus. Under service-oriented arrangements, the service provider is responsible for the full spectrum of support, including ownership, sustainment, and operation of assets. Furthermore, contracting arrangements will include incentives and penalties against levels of support service or delivery. The service provider will need to think differently and design the output solutions that deliver the desired outputs as well as generating profit. This is a different type of business with unfamiliar contractual metrics and risks. In PSS, the emphasis is on customisation of solution designs to meet service needs and create new values of use for the customer [23].

A performance-based contract in PSS will include incentives and penalties against levels of support service or delivery as discussed before. Hence, the service contractor should have a thorough understanding of how the system works and how the supporting systems around the asset provide the services to achieve the desirable performance [24]. Due to this highly individualised nature of service, no one performance-based contract is the same. The support system then becomes a one-off development which imposes significant system design issues to both asset owners and contractors.

A service contract often involves active interaction within the supply chain and with the customer. In a service environment, it is normal that a new separate enterprise is formed from several independent, collaborating enterprises. There are many risks in this strategy, for example, there are risks in collaboration, confidentiality, intellectual property, transfer of goods, conflicts, opportunity loss, product liability, and others. To minimise the risks for the new service enterprise, enterprise engineering researches provide an enterprise architecture framework as a common starting point. The study of enterprise architecture in the last decade has been on how enterprises can be designed and operated in an environment when the enterprise missions and objectives are clear. They assumed that one can follow the common engineering practice of well established sequences of steps: design, implementation, operation, and decommission phases [25]. The rationale to use enterprise engineering methodologies to guide these steps is to minimize enterprise design modifications and associated rework of the system governing information and material flows [26]. A PSS as discussed in this paper is a dynamic system. Any unplanned change to the enterprise is an impact of uncertainty to enterprise performances. The enterprise architecture (EA) approach provides a structured system to manage services activities, for example, promote planning, reduce risk, implement new standard operating procedures and controls, and rationalize manufacturing facilities [27]. Hence, this paper uses an EA approach to understand the enterprise under which the product and related services are managed and to assess the performance of the PSS.

ARCHITECTURAL APPROACH

An enterprise architecture defines methods and tools which are needed to identify and carry out change [28]. Enterprises need lifecycle architecture that describes the progression of an enterprise from the point of realisation that change is necessary through setting up a project for implementation of the change process. Denton et al. [29] specified an information technology route map that enabled rapid design of IT solutions to automate some business processes for service supply chains. Therefore, it is crucial to use a systematic design methodology that helps the management developing well-defined policy and process across the organisational boundaries and implements the changes in all enterprises of the service supply chain.

However, traditional enterprise architectures are based on top down approach. They emphasized on uniformity throughout the organization. As such, the structure is inflexible. Changing the structure in order to respond to fast changing dynamic issues for in service engineering systems will be too long to fix any problem [30]. Two issues in using standard enterprise architecture to model service and support systems are identified.

Inwards Modelling

Existing enterprise architectures contain functions, data, staff, resources which are inwards looking and focus on internal company issues. There is very few, if any, modelling constructs for interaction with other systems.

Static, Snapshot View of Present and Future

EA modelling methodology is based on the understanding that it is a snapshot of the enterprise at certain point in time. Service systems are dynamic organisations. There are frequent staff movements, external environment changes, customer changes, and change of use context. The static nature of existing EA is incapable of handling changes as anticipated in real service systems.

In order to support decisions on business opportunities by enterprise architectural approach, the PSS enterprise should have the following characteristics.

Measurement of Performance and the Development of Metrics That Can Be Supported by Technology

The PSS will be operated in parallel with the complex engineering system. Service is qualitatively different to the familiar product-based approach where hard artefacts are delivered to the asset owner. Service is a negotiated exchange with the asset owner (and operator) to provide intangible outputs that are usually produced together with the asset owner. A service is usually consumed at the time of production. Services cannot be transferred to other asset owners in the same way that products can. Hence, the development of appropriate performance metrics is essential and most of these are supported by advanced information and computational technologies.

Use of Proven Enterprise Architecture That Incorporates Broad Range of Engineering Disciplines

PSS incorporates system design knowledge that draws upon principles derived from a wide range of engineering disciplines including systems engineering, logistics engineering, project management, information systems, and many others. The knowledge helps the system support engineer to take into account as many constraints as possible during the system design phase. These constraints are imposed by the environment in which the complex system and the business are operating.

Sustainability Capability That Manages Risks in the Support Contract

The performance-based services are characterised by the need to create value for both asset owner and the service provider. As such both sides are treated as coinnovators in the design of the PSS. Many decisions are made based on incomplete data rather than fully analysed data set. There are a lot of risks, both from the point of view of data availability, as well as subjective human judgement and communication.

When customers want to outsource a service function, capturing the requirements is the real challenge for human intelligence and ability to manage what we know, what other people know, and what nobody knows [31]. A modelling construct that has more human interaction characteristics is required.

SHEL model has been developed from analysing and modelling human interaction with physical and project activities [32]. Chang and Yeh [33] applied the SHEL model to describe the structure of the air traffic control system and its interface to human operators. The research findings provided practical insights in managing human performance interfaces of the system due to changes in its operating environment. Felici et al. [34] applied the SHEL model to deal with the definition of the requirements for a new railways traffic control system. Lei and Le [35] evaluated risks of human factors in flight deck system. They used a SHEL model and found five most significant factors on the risks in the system.

Extending from traditional enterprise modelling methodology, Chattopadhyay and Mo [36] modelled a global engineering services company as a three-column progression process that was centred on human engineering effort. Chattopadhyay et al. [37] developed a business model for virtual manufacturing with particularly emphasis on the need for intense collaborative network for a variable-variety, variable-volume and manufacture to order situation with provisions for recycling and reverse logistics. The concept was further developed as an aggregated model resembling nature's atomic and molecular interaction after studying the supply chain in China [38]. These new attempts to incorporate human participation in modern global enterprises have highlighted the effect of new information and communication technologies in bringing the human dimension in enterprise architecture to a dominated position.

As seen from the literature, the SHEL model has particular focus on local, operational level of the enterprise. It does not have the support of engineering methodology to ensure repeatability and sustainability of the system. Likewise, traditional enterprise architecture methodology tends to ignore human interaction and becomes difficult to describe vibrant enterprises in the services sector. It is logical to develop a new enterprise model for services that combines traditional enterprise architecture with SHEL concept. We propose this new complex engineering product service system architecture as shown in Figure 2.

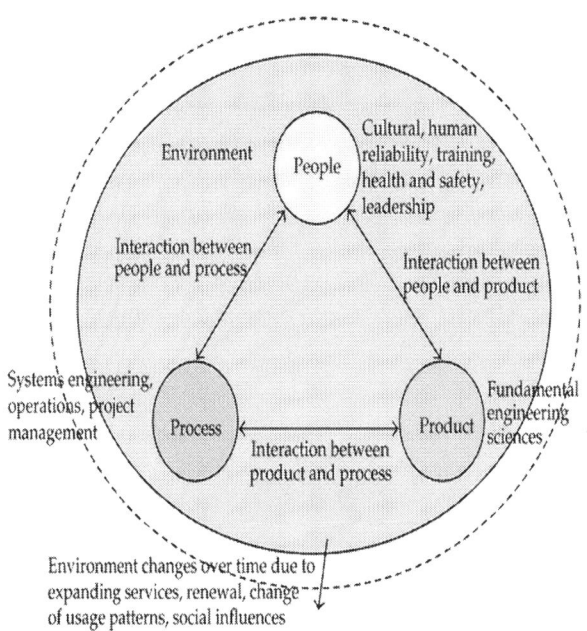

Figure 2: Complex engineering product service system (CEPSS) architecture.

Figure 2 shows that a PSS for complex engineering products should be a four–dimensional system architecture: product, process, people, and environment. This is in contrast to conventional enterprise modelling methodologies that had significant influence to system development thinking in the 1990s. The new architecture covers the additional "changing" aspect of service system by integrating the concepts of product, process, people to changes in environment over time. The four dimensions are interlinked and affecting one another. The architecture provides a focus for consolidating existing knowledge of designing a service system as well as an instrument for projecting future requirements in a system so that new features can be developed in an orderly fashion.

CASE STUDY RESEARCH DESIGN

Case study research is particularly useful in identifying specific characteristics that affect system performances. Serra and Ferriera [39] identified four strategy pillars in five case studies of well-known multinational corporations. In supply chain case study research, Seuring [40] surveyed 68 papers related to supply chain sustainability and supply chain performance and concluded that more supply chain cases should be documented and reviewed. Lewis et al. [41] researched three case studies in the energy and maintenance management practices. They found that the link between different service requirements should be better understood when the teams worked together for service solutions.

However, extracting the theoretical essence of a PSS is not a trivial exercise from studying a wide variety of cases. For example, Holschbach and Hofmann [42] used case study evidence from eight manufacturing and eight service companies. They found that companies did not use quality management for externally sourced business services to its full potential. There were major difficulties in determining quality failures, standardization, and quantity of service. Zhang et al. [43] carried out a structured literature review on the influence of ICT in supply chain management. They found that despite inconsistency in reported findings in this field of research, there were general positive performance outcomes of supply chains due to ICT system development. Kucza and Gebauer [44] investigated the forms of servitization of products could help global manufacturing firms to develop new service-based and relationship-based value propositions for customers. Four such forms were identified: integrated and ethnocentric; integrated and polycentric; separated and polycentric; and separated and geocentric.

One of the most difficult issues in the case study research is the definition of unit of analysis. Grünbaum [45] provided a useful elaboration of the concept by examples of generic case studies based on modifications of Yin's [46] case study design. Huang [47] interviewed top management of 40 SMEs in Taiwan on their perceptions of IT components in business and found five internal strategic factors inhibiting top management support in IT adoption. M. Bielli and A. Bielli [48] presented a conceptual model of a SMEs network in the European project CO-DESNET. The coordination of distributed and autonomous agents characteristics of the collaborative enterprise clusters was represented by suitable models such that global performances could be evaluated. The model was validated by a case study outlining a transition to the Net Economy of SMEs in Italy.

In this paper, three case studies are described and their key features are highlighted. The cases are earlier forms of PSS representing various degrees of success in creating new businesses for the equipment suppliers. The ad hoc systems established at the time of these cases provide good examples for benchmarking current thinking of the design and implementation of PSS. These cases are chosen because the parties in the cases have tried to apply a defined enterprise infrastructure that links different parts of the service system working in conjunction with the product. Subsequently, the service system has to be re-designed and tailored to characteristics of the product or the enterprise itself.

Data collection for this type of case study research depends on the relationship of the researchers and the parties in the cases. In all cases, the author of this paper has had varying levels of participation in the cases.

CASE STUDIES

The products in the three cases are complex engineering systems. Case 1 is a computer controlled plasma cutting machine that can cut steel plates up to 50 mm thick. The machine has been sold over the world. Case 2 is a chemical plant that is designed and built by a Japanese engineering company. In order to support the customer with minimum costs, the support system was designed to use the Internet, which was evolving at the time when the project was done. Case 3 is a defence case in which the ship builder formed a service consortium to continue its business after the ships were built. Evaluation and analysis of the cases will be based on the CEPSS model presented in Figure 2

Case Study 1: Signal-Based Condition Monitoring System

System health monitoring plays a critical role in preventative maintenance and product quality control of modern complex engineering products.

The effectiveness of management can directly impact their efficiency and cost-effectiveness. A condition monitoring system monitors the products using various classical methods of signal analysis such as spectrum or state-space analyses [49]. Maintenance decisions are then made according to the prediction of system performance.

Using time-based signals available from normal machine sensing mechanisms, a CNC machine manufacturer in Australia developed a remote condition monitoring system for plasma CNC cutting systems with the aim of servicing the customer anywhere in the world via the Internet. Figure 3 shows the network structure of the system known as ROSDAM [50]. All ROSDAM-enabled machines were configured as servers that had functionality communicating with the global master server. Information about the operation of the machines was captured through individual companies' database. The significantly improved sources of information enabled the product manufacturer to decide the best option that supported operation and maintenance of the plasma cutting machine from a distance.

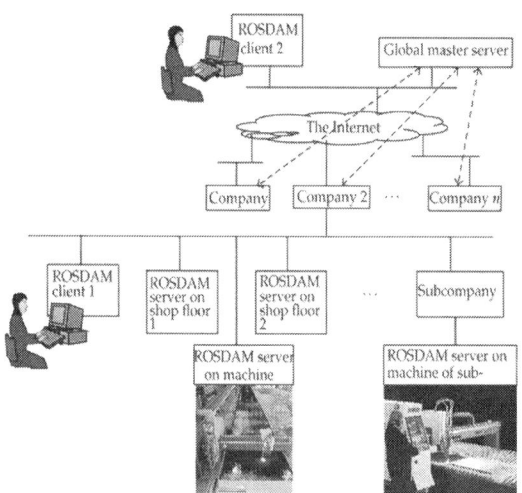

Figure 3: Signal-based condition monitoring service system network structure.

In this case study, new elements were required to be developed and integrated with the product, that is, the CNC plasma cutting machine. These elements are mapped to the CEPSS model as shown in Table 1.

Table 1: Mapping of service elements in case 1 to CEPSS

Element	Description	Mapped to CEPSS
On machine signal-based diagnostics capability	A new diagnostics software module based on chaotic theory and digital signal processing was developed to assist identification of faults.	Product
Communication networks and IT systems based on client-server model	The controller of the machine was significantly changed from a normal standalone operating system to one that can act as a server in a network environment.	Product-People
Knowledge sharing—transform customer data to information to knowledge	New data processing algorithms were developed as software modules that were required to process data on machine to knowledge useful to enhance operational efficiency.	People

Engineering information integrated for supporting more effective customer service	Engineering information such as bill of material, machine configuration management, parts inventory, and resources planning were integrated from different sources including CAD, MRP, and various manufacturing sources to create seamless operation database for the machine.	Process
The new system design requires upgrade of field products	Field upgrade for machines that were already installed at customers' location was progressively rolled out according to contracted maintenance schedules.	Product-Environment

Case Study 2: Global Operation Support Services

Complex assets are normally built from a large number of components and involving a large number of engineers and contractors. In the past, customers as plant owners usually maintain their own service department. However, the increasing complexity of the plant and operating conditions such as environmental considerations require service personnel to have a higher level of analysis and judgment capability. Rathwell and Williams [51] used Flour Daniel as the study platform and validated the use of enterprise engineering methodology for creating services that support operations of chemical plant. The study showed that significant efficiency gain could be achieved in the design and implementation of the service system through systematic enterprise modelling analysis.

In managing the design and manufacture of a chemical plant for their customer, Kamio et al. [52] established a service virtual enterprise (SVE) with several partner companies around the world providing after-sales services to a customer (Figure 4).

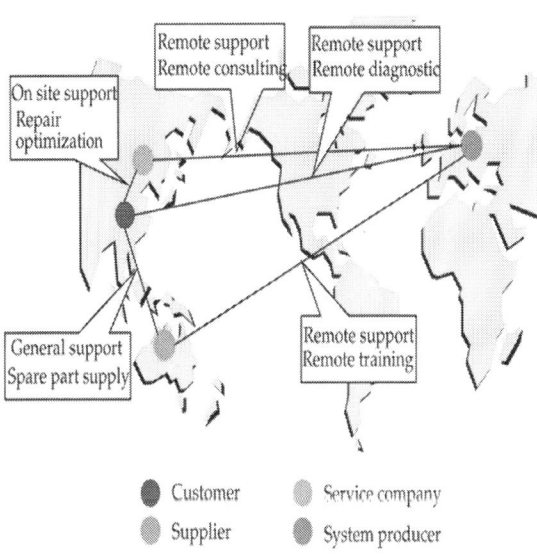

Figure 4: A global service virtual enterprise.

Each partner in Figure 4 was an independent entity that was equipped with its own unique capabilities and competencies, assuming responsibility to perform the allocated work. The SVE was designed as a "hosting service" which had a broad range of services including plant monitoring, preventive maintenance, trouble-shooting, performance simulation and evaluation, operator training, knowledge management, and risk assessment. Participants of the virtual enterprise had well-defined roles and responsibilities.

An essential element in the design of a service enterprise is to develop efficient system architecture and provide the right resources to the right service tasks. By synchronising organisational activities, sharing information and reciprocating one another's the technologies and tools, each partner in the service enterprise will be able to provide services that would have been impossible by individual effort. The PSS therefore requires properly designed components to support the use of technology in the provision of support services to customers.

In this case study, in order to operate the SVE, new service elements were developed. They are listed and mapped to CEPSS in Table 2.

Table 2: Mapping of service elements in SVE to CEPSS

Element	Description	Mapped to CEPSS
Intercompany communication networks and IT systems	New IT and communication systems were installed to enable intercompany exchange of information as well as personal interaction.	Process-People
Work items synchronized within the project across companies	Work items were analysed individually so that the link from individual level to group level can be streamlined ensuring minimum duplication of work and conflicts.	Process
Change of human organisation role	The relationship within a SVE was definitely different from a totally authoritative company structure. A much more flexible human organisation structure was established.	People

Global access by customer	The SVE was implemented on the Internet. At the time of the development of this PSS, the Internet was still not entirely effective in some parts of the world, and this environmental constraint imposed significant challenges on the SVE development.	Process-Environment

It should be noted that the engineering product remained the same as it was designed initially. There was no noticeable engineering change required on the product itself in order to implement the support service offered by SVE.

Case Study 3: Ship Service System

The ANZAC Ship Alliance (ASA) could be thought of as a virtual company with shareholders comprising the Australian government and two commercial companies, one of which was the ship builder. The primary goal was to create best value for money [53]. The primary goal of ASA was to manage all changes and upgrades to the ANZAC Ships [54]. The Alliance was a "solution focused" company where the staff of the ANZAC Ship Alliance Management Office would develop change solutions but the detailed design is undertaken by the "shareholders" drawing upon their existing and substantial knowledge of the ANZAC Class.

Prior to the development of ASA, Hall [55] developed a highly integrated documentation and configuration management system that served the on-going need of ten ANZAC class frigates. Over the life time of the asset (30 years), changes due to new technologies, people and defence requirements are inevitable. The organisation structure of ASA can be described as shown in Figure 5.

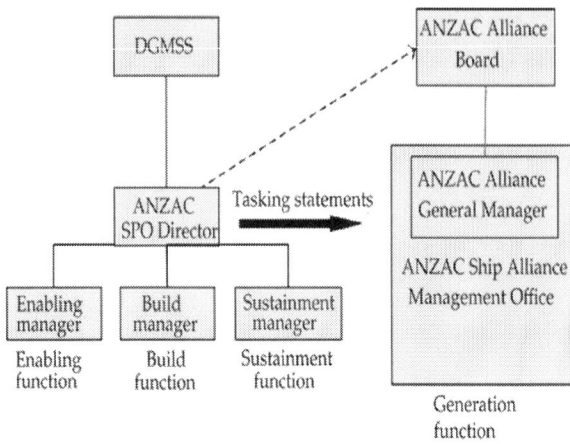

Figure 5: Organisation structure of ANZAC Ship Alliance.

In this case study, the enterprise was not set up as a legal entity. There was no formal binding agreement among the partners in ASA. In the language of virtual enterprise, the partners were loosely linked organisations such that everything done in the ASA is based on trust. The new service elements that were developed on this premise were listed in Table 3.

Table 3: Mapping of service elements in ASA to CEPSS

Element	Description	Mapped to CEPSS
New procedures and legal processes	All parties joined the ASA with the understanding of a set of business rules. (i) All parties win or all parties lose (ii) Collective responsibility, equitable sharing of risk and reward (iii) All decisions based on "best for project" philosophy (iv) Access to resources, skills, and expertise of all parties (v) All financial transactions are fully open book (vi) Encouragement of innovative thinking—outstanding outcomesThese nontraditional business rules imposed challenges on the ASA as the service system of the ships.	Process

Ownership of product-related services rested with ASA	The enterprise was a joint development of several companies and hence the ownership of the product-related services had to be resolved. This issue settled at the end but as all participants were in the defence business environment, and the customer was a partner of ASA, the "company" structure at the right hand side of Figure 5 was able to provide sufficient background understanding for the people to rely on.	Process-Product
Secondment of staff from the three partner organisations	As the "company" status was eventually accepted, the need to develop a set of processes that is acceptable to all staff (who were all seconded from the partner organisations) became urgent. A lot of time was spent in synchronising practices and culture originating from individual companies. There was confusion in the first year among the staff of the participating organization about the nature of ASA. This issue was resolved through a number of ASA workshops.	People-Environment

From the point of view of the ship builder company, the ASA was an unprecedent business environment that no one knew exactly how to operate. There were some upgrade projects as continuous support initially. After several years of operation, the ASA entered into a new material support program focusing on supplies and shore facilities.

PERFORMANCE-CAPABILITY ASSESSMENT

From the foregoing case studies, the subsystems and interactions between the subsystems of the CEPSS model are represented by specific elements in the cases. Table 4 summarises the relevance of the cases in a matrix.

Table 4: Applicability of cases to elements and interactions of CEPSS

	Product	**People**	**Process**
Product	Case 1		
People	Cases 1 and 2	Cases 1,2,3	
Process	Case 3	Case 2	Cases 1,2,3

When a PSS enterprise is created by three interdependent subsystems under the CEPSS model, the ability of the total PSS in meeting the performance expectation of the customer will depend on how the capability of each of the subsystems is designed and the effect of the environment on the execution of these subsystem capabilities. Theoretically, for each of the subsystems in a CEPSS enterprise, it is possible to devise some measures of enterprise capability in relation to the outcomes that can be produced by the capability. These methods include survey, interviews, system audit, comparative analysis, human resources records, and so forth. This type of capability assessments are bound to have certain degree of uncertainty. In addition, the enterprise capability will change over time due to changes in people, process, and product.

Likewise, by way of aggregation of subsystem capabilities, the capability of the PSS can be benchmarked against the theoretical capabilities required to achieve expected performance, an assessment of the potential achievement can be made at the outset, when the PSS enterprise is established. Due to the uncertainties as explained above, the probability of the PSS enterprise's achievement can be expressed in terms of the frequency of success of this capability meeting specified performance metrics.

Using case 1 as an example, assuming all other capabilities are able to deliver to the required performance standards, the improved service elements can be assessed using a 5-point scale of 1 to 5 where 5 represents most certain and 1 represents rarely meeting expectation as shown in Table 5.

Table 5

CEPSS	Element	Customer expectation	Assessed capability	Rating
Product	On machine signal-based diagnostics capability	Diagnose the causes for all (100%) of reported problems.	The new signal diagnostics capability only provides high level fault analysis.	3
Product-People	Communication networks and IT systems based on client-server model	Obtain online support information at any time.	Connection to the master server depends on the Internet	5
People	Knowledge sharing— transform customer data to information to knowledge	Receive correct advice all the time.	Fault trees based on past records sometimes contain conflicting cases	4
Process	Engineering information integrated for supporting more effective customer service	Identify suitable machine configuration and spare parts.	Support system is integrated with product design database which requires minor interface development.	4
Product-Environment	The new system design requires upgrade of field products	Upgrade as minimum cost and within reasonable time.	Establish a separate group within the company responsible for upgrade requests.	5

The ratings in Table 5 are for illustration only. Since most ratings are above average, the PSS in case 1 is assessed as likely to meet customer expectation. With an assessment of the capability against expectation outcomes, the probability of the service contract being successful can then be determined from the contractual terms.

As an illustration, if the Defence Materiel Organisation four-band performance incentive/penalty scheme is used, the capability distribution can be overlaid on the achieved performance axis as shown in Figure 6. The capability distribution (in dotted curve) represents the probability density of the enterprise achieving a performance level on the x-axis. Different risks and probabilities of a PSS contract can then be identified as shown in Figure 6.

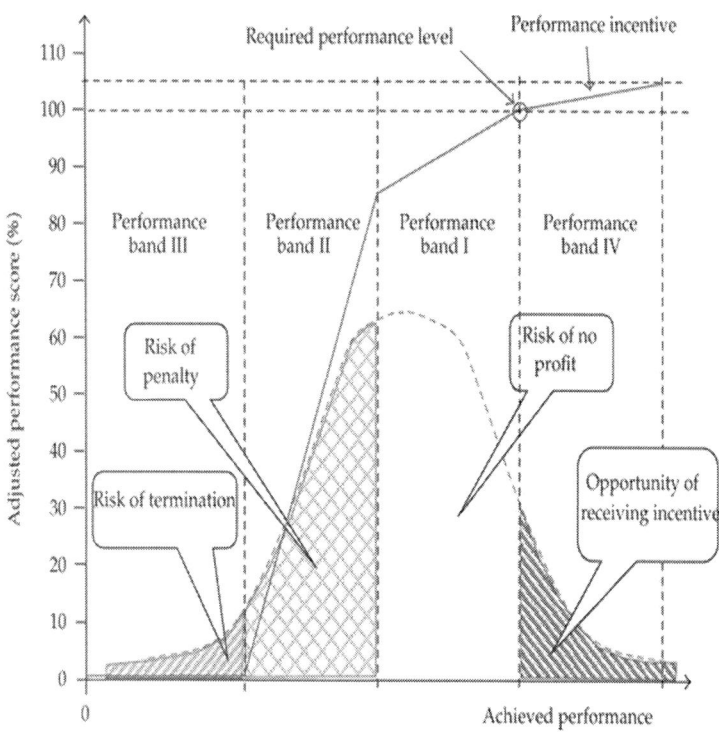

Figure 6: Performance band and risks in the PSS contract.

Several decisions can be made using this assessment outcome. For this discussion, the CEPSS contractor is the prime contractor who is a major engineering company working with the client on a new complex engineering system within the capability systems lifecycle shown in Figure 1. Using the performance-capability assessment methodology, the CEPSS contractor has visibility on what risks are likely to incur in its current proposal.

First, based on this information, the CEPSS contractor can decide whether to go ahead with the enterprise capabilities he/she has. This is a go and no-go decision scenario. The contractor will have to decide in conjunction with other concurrent opportunities, which may be assessed by the same PSS opportunity assessment methodology or other means.

Second, if the risk level is too high, the CEPSS contractor can increase his/her enterprise capabilities by raising the contract price to cover the costs, or by implementing organisational improvements such as lean and six sigma. In the latter case, the time factor of the CEPSS will be brought in to map the change of capability over time.

Third, the CEPSS contractor can identify the shortfall capability areas and collaborate with other prime suppliers in the industry. The performance capability assessment is then modified as the overall performance capability of the combined CEPSS consortium. The contractual detail of the coalition arrangement is outside the scope of this paper [56]. A mapping of potential changes over time in capabilities of other parties should also be considered, as highlighted by the CEPSS model.

Fourth, the CEPSS contractor can consider boosting its core capabilities by mergers and acquisition with other companies. This case is more complicated since consideration of which companies to acquire depends on strategic alignment requirements. However, this option will represent an immediate shift of the capability distribution to the right. The only concern is whether the new organisation can be restructured and operated effectively and quickly enough for executing the PSS contract [57].

CONCLUSIONS

New business models in delivering capabilities from the operations of complex engineering products such as aircraft, ships, and refineries have favoured the integration of acquisition and sustainment phases of the products. The product service system (PSS) concept enables manufacturers of these complex engineering products to incorporate support services into the product's system capability lifecycle. These services are substantially more complex than routine, reliability-based maintenance or spare parts support. Unfortunately, in the past decade, researches in the development of support systems have been fragmented. There is no unified body of knowledge specifying the methodologies that can naturally lead to the design of a support solution for any scenario. This situation prompted this study.

The new type of service business model, which is represented by performance-based contracts, focuses on the performance of the complex engineering product during operations in terms of timeliness, availability, maintainability, and sustainability costs in the product's complete lifecycle from conception, design, manufacture to disposal. Ultimately, the service business model is expected to provide long-term benefits to both contractor and customer due to efficiency gains. However, the PSS itself has imposed significant risks to the contractor not only in the manufacture of the product, but also in the provision of support services over long period of time at a fixed reward scheme with a lot of unknowns. A successful PSS enterprise requires an analysis methodology that can assist the contractor to estimate the performance outcomes of the PSS.

In this paper, three case studies are analysed using case study research design approach to investigate a new complex engineering product service system (CEPSS). CEPSS combines the conceptual elements in the SHEL model with the systematic enterprise architecture modelling approach. Service elements of the three case studies have been mapped to CEPSS. Through this mapping, the CEPSS can be broken down into subsystems. Capabilities of the subsystems are readily assessed against customer expectation using qualitative methods such as opinion surveys. Once the capabilities are known, they can be overlaid on the performance-based incentive/penalty scheme so that different risk levels can be assessed as a decision support tool. Decision makers can then use this information to select options related to the business opportunities presented to them.

This paper is a preliminary investigation of the fundamental question: what service elements should be developed in the new PSS environment? Investigation using case study approach so far seems to show that the capability of the PSS can be assessed by an aggregated evaluation of these service elements and hence the expected performance of the service system can be estimated. However, the complexity of the system cannot be ignored. Further research is required to create a consistent scoring framework that can be applied across different risks and engineering systems. Naturally, more case studies on a broad range of engineer products would be necessary, while validating of the CEPSS using a more quantitative, evidence-based approach against these cases is vital to the development of a quantitative consistent scoring framework.

REFERENCES

1. Defence Materiel Organisation, Performance Based Contracting Handbook—Guiding Principles and Performance Framework, Version 2. 0, Australian Department of Defence, 2007,http://www.defence.gov.au/dmo/asd/publications/asd_pbc_v2.pdf.

2. Ministry of Defence, "Defence Industrial Strategy," Defence White Paper, The UK Defence White Paper, 2005.

3. S. Brammer and H. Walker, "Sustainable procurement in the public sector: an international comparative study," International Journal of Operations & Production Management, vol. 31, no. 4, pp. 452–476, 2011.

4. H. Corsten and R. Gössinger, "Output flexibility of service enterprises—an analysis based on production theory," International Journal of Production Economics, vol. 104, no. 2, pp. 296–307, 2006.

5. D. J. A. Sherwin, "A review of overall models for maintenance management," Journal of Quality in Maintenance Engineering, vol. 6, no. 3, pp. 138–164, 2000. ·

6. F. T. S. Chan, H. C. W. Lau, R. W. L. Ip, H. K. Chan, and S. Kong, "Implementation of total productive maintenance: a case study," International Journal of Production Economics, vol. 95, no. 1, pp. 71–94, 2005.

7. A. S. B. Tam, W. M. Chan, and J. W. H. Price, "Optimal maintenance intervals for a multi-component system," Production Planning and Control, vol. 17, no. 8, pp. 769–779, 2006.

8. G. Abdul-Nour, M. Demers, and R. Vaillancourt, "Probabilistic safety assessment and reliability based maintenance policies: application to the emergency diesel generators of a nuclear power plant,"Computers & Industrial Engineering, vol. 42, no. 2–4, pp. 433–438, 2002.

9. S. Colombo and M. Demichela, "The systematic integration of human factors into safety analyses: an integrated engineering approach," Reliability Engineering and System Safety, vol. 93, no. 12, pp. 1911–1921, 2008.

10. Y. Shen, "Selection incentives in a performance-based contracting system," Health Services Research, vol. 38, no. 2, pp. 535–552, 2003.

11. C. J. Heinrich and Y. Choi, "Performance-based contracting in social welfare programs," The American Review of Public Administration, vol. 37, no. 4, pp. 409–435, 2007.

12. K. A. Chatha, J. O. Ajaefobi, and R. H. Weston, "Enriched multi-process modelling in support of the life cycle engineering of Business Processes," International Journal of Production Research, vol. 45, no. 1, pp. 103–141, 2007.

13. Defence Materiel Organisation, ASDEFCON (Support) V3. 0 Performance Management Framework, Version 3. 0, Australian Department of Defence, 2011,http://www.defence.gov.au/dmo/gc/asdefcon/asdefcon_support/vers_3/PPBC_Framework.pdf.

14. I. C. L. Ng, G. Parry, D. McFarlane, and P. Tasker, "Towards a core integrative framework for complex engineering service systems," in Complex Service Systems: Concepts and Research, I. C. L. Ng, P. Wild, G. Parry, D. McFarlane, and P. Tasker, Eds., Springer, London, UK, 2011.

15. T. Abe, "What is service science," Research Report 246, The Fujitsu Research Institute, Economic Research Centre, Tokyo, Japan, 2005.

16. A. Tukker, "Eight types of product-service system: eight ways to sustainability? Experiences from suspronet," Business Strategy and the Environment, vol. 13, no. 4, pp. 246–260, 2004.

17. T. S. Baines, H. W. Lightfoot, S. Evans et al., "State-of-the-art in product-service systems," Journal of Engineering Manufacture, vol. 221, no. 10, pp. 1543–1552, 2007.

18. J. Spohrer, P. P. Maglio, J. Bailey, and D. Gruhl, "Steps toward a science of service systems,"Computer, vol. 40, no. 1, pp. 71–77, 2007.

19. P. Johansson and J. Olhager, "Linking product-process matrices for manufacturing and industrial service operations," International Journal of Production Economics, vol. 104, no. 2, pp. 615–624, 2006.

20. A. D. Neely, "Exploring the financial consequences of the servitization of manufacturing," Operations Management Research, vol. 1, no. 2, pp. 103–118, 2009. ·

21. Z. J. M. Shen and M. S. Daskin, "Trade-offs between customer service and cost in integrated supply chain design," Manufacturing and Service Operations Management, vol. 7, no. 3, pp. 188–207, 2005.

22. IfM and IBM, "Succeeding through Service Innovation," A Discussion Paper, University of Cambridge, Cambridge, UK, 2007.

23. I. C. L. Ng, G. Parry, L. Smith, and R. Maull, "Value co-creation in complex engineering service systems: conceptual foundations," in Proceedings of the Forum on Markets and Marketing: Extending the Service Dominant Logic, pp. 24–26, Cambridge, UK, September 2010.

24. R. D. Behn and P. A. Kant, "Strategies for avoiding the pitfalls of performance contracting," Public Productivity and Management Review, vol. 22, no. 4, pp. 470–489, 1999.

25. G. Doucet, J. Gøtze, P. Saha, and S. Bernard, "Coherency management: using enterprise architecture for alignment, agility, and assurance," Journal of Enterprise Architecture, vol. 4, no. 2, pp. 9–20, 2008.

26. V. Veneziano, S. Jones, and C. Britton, "Adding a systemic view to the requirements engineering processes," in Proceedings of the 9th International Workshop on Database and Expert Systems Applications, pp. 321–325, Florence, Italy, September 1999.

27. H. Shen, B. Wall, M. Zaremba, Y. Chen, and J. Browne, "Integration of business modelling methods for enterprise information system analysis and user requirements gathering," Computers in Industry, vol. 54, no. 3, pp. 307–323, 2004.

28. P. Bernus and L. Nemes, "A framework to define a generic enterprise reference architecture and methodology," Computer Integrated Manufacturing Systems, vol. 9, no. 3, pp. 179–191, 1996.

29. P. D. Denton, D. Little, R. H. Weston, and A. Guerrero, "An enterprise engineering approach for supply chain systems design

and implementation," International Journal of Services and Operations Management, vol. 3, no. 2, pp. 131–151, 2007.

30. J. P. T. Mo and L. Nemes, "Issues using EA for merger and acquisition," in Coherency Management: Architecting the Enterprise For Alignment, Agility, and Assurance, G. Doucet, J. G. Gøtze, P. Saha, and S. Bernard, Eds., Chapter 9, pp. 235–262, Author House, 2010.

31. D. Kwon, W. Oh, and S. Jeon, "Broken ties: the impact of organizational restructuring on the stability of information-processing networks," Journal of Management Information Systems, vol. 24, no. 1, pp. 201–231, 2007.

32. G. J. Molloy and C. A. O›Boyle, "The SHEL model: a useful tool for analyzing and teaching the contribution of human factors to medical error," Academic Medicine, vol. 80, no. 2, pp. 152–155, 2005.

33. Y. H. Chang and C. H. Yeh, "Human performance interfaces in air traffic control," Applied Ergonomics, vol. 41, no. 1, pp. 123–129, 2010.

34. M. Felici, M. A. Sujan, and M. Wimmer, "Integration of functional, cognitive and quality requirements. A railways case study," Information and Software Technology, vol. 42, no. 14, pp. 993–1000, 2000.

35. W. Lei and D. Le, "Risk evaluation of human factors in flight deck system," in Proceedings of the 2nd IEEE International Conference on Advanced Management Science (ICAMS ‹10), pp. 381–385, Chengdu, China, July 2010.

36. S. Chattopadhyay and J. P. T. Mo, "Modelling a global EPCM (Engineering, Procurement and Construction Management) enterprise," International Journal of Engineering Business Management, vol. 2, no. 1, pp. 1–8, 2010.

37. S. Chattopadhyay, D. S. K. Chan, and J. P. T. Mo, "Business model for virtual manufacturing: a human-centred and eco-friendly approach," International Journal of Enterprise Network Management, vol. 4, no. 1, pp. 39–58, 2010.

38. S. Chattopadhyay, D. S. K. Chan, and J. P. T. Mo, "Modelling the disaggregated value chain—the new trend in China," International

Journal of Value Chain Management, vol. 6, no. 1, pp. 47–60, 2012.

39. F. R. Serra and M. P. Ferreira, "Emerging determinants of firm performance: a case study research examining the strategy pillars from a resource-based view," Management Research, vol. 8, no. 1, pp. 7–24, 2010.

40. S. A. Seuring, "Assessing the rigor of case study research in supply chain management," Supply Chain Management, vol. 13, no. 2, pp. 128–137, 2008.

41. A. Lewis, A. Elmualim, and D. Riley, "Linking energy and maintenance management for sustainability through three American case studies," Facilities, vol. 29, no. 5, pp. 243–254, 2011. ·

42. E. Holschbach and E. Hofmann, "Exploring quality management for business services from a buyer›s perspective using multiple case study evidence," International Journal of Operations & Production Management, vol. 31, no. 6, pp. 648–685, 2011.

43. X. Zhang, D. P. van Donk, and T. van der Vaart, "Does ICT influence supply chain management and performance?: a review of survey-based research," International Journal of Operations & Production Management, vol. 31, no. 11, pp. 1215–1247, 2011.

44. G. Kucza and H. Gebauer, "Global approaches to the service business in manufacturing companies,"Journal of Business & Industrial Marketing, vol. 26, no. 7, pp. 472–483, 2011.

45. N. N. Grünbaum, "Identification of ambiguity in the case study research typology: What is a unit of analysis?" Qualitative Market Research, vol. 10, no. 1, pp. 78–97, 2007.

46. R. K. Yin, "Case study methods," in Complementary Methods for Research in Education, J. L. Green, G. Camilli, P. B. Elmore, A. Skukauskaite, and E. Grace, Eds., Chapter 6, pp. 111–122, American Education Research Association, Washington, DC, USA, 3rd edition, 2006.

47. L. K. Huang, "Top management support and IT adoption in the Taiwanese small and medium enterprises: a strategic view," International Journal of Enterprise Network Management, vol. 2, no. 3, pp. 227–247, 2008.

48. M. Bielli and A. Bielli, "Innovation paths management in

collaborative enterprise clusters,"International Journal of Enterprise Network Management, vol. 2, no. 4, pp. 366–376, 2008.

49. S. Yang, M. Sammut, T. Kearney, and J. P. T. Mo, "Engine condition monitoring with ignition signals," in Proceedings of the Intelligent Vehicle & Road Infrastructure Conference, Melbourne, Australia, February 2005.

50. J. P. T. Mo, "Case study—farley remote operations support system," in Enterprise Integration Handbook, P. Bernus, L. Nemes, and G. Schmidt, Eds., Chapter 21, pp. 739–756, Springer, New York, NY, USA, 2003.

51. G. A. Rathwell and T. J. Williams, "Use of the purdue enterprise reference architecture and methodology in industry," in Model and Methodologies For Enterprise Integration, pp. 12–44, Chapman & Hall, London, UK, 1996.

52. Y. Kamio, F. Kasai, T. Kimura, Y. Fukuda, I. Hartel, and M. Zhou, "Providing remote plant maintenance support through a service virtual enterprise," in Global Engineering and Manufacturing in Enterprise Networks, vol. 224, pp. 195–206, VTT Symposium, Helsinki, Finland, December 2002.

53. C. Clifton and C. F. Duffield, "Improved PFI/PPP service outcomes through the integration of Alliance principles," International Journal of Project Management, vol. 24, no. 7, pp. 573–586, 2006.

54. J. P. T. Mo, M. Zhou, J. Anticev, L. Nemes, M. Jones, and W. P. Hall, "A study on the logistics and performance of a real virtual enterprise," International Journal of Business Performance Management, vol. 8, no. 2-3, pp. 152–169, 2006.

55. W. P. Hall, "Managing technical documentation for large defence projects: engineering corporate knowledge," in Proceedings of the Global Engineering, Manufacturing and Enterprise Networks, pp. 370–378, Melbourne, Australia, November 2000.

56. L. Nemes and J. P. T. Mo, "Collaborative networks in Australia—challenges and Recommendations," in Collaborative Networked Organizations, L. M. Camarinha-Matos and H. Afsarmanesh, Eds., pp. 97–102, Kluwer Academic, New York, NY, USA, 2004.

57. Y. Xiaoli and M. Shanley, "Industry determinants of the "merger versus alliance" decision," Academy of Management Review, vol. 33, no. 2, pp. 473–491, 2008.

Chapter 3

Thermodynamic Analysis and Synthesis Gas Generation by Chemical-Looping Gasification of Biomass with Nature Hematite as Oxygen Carriers

Zhen Huang, Fang He, Anqing Zheng, Kun Zhao, Sheng Chang, Xinai Li, Haibin Li, and Zengli Zhao

Key Laboratory of Renewable Energy and Gas Hydrate of Chinese Academy of Sciences, Guangzhou Institute of Energy Conversion, Chinese Academy of Sciences, Guangzhou, China

ABSTRACT

Thermodynamic parameters of chemical reactions in the system were carried out through thermodynamic analysis. According to the Gibbs free energy minimization principle of the system, equilibrium composition of the reactions of chemical-looping gasification (CLG) of biomass with natural hematite (Fe_2O_3) as oxygen carrier were analyzed using commercial software of HSC Chemistry 5.1. The feasibility of the CLG of biomass with hematite was experimental verified in a lab-scale bubbling fluidized bed reactor using argon as fluidizing gas. It was indicated the experimental results were consistent with the theoretical analysis. The presence of oxygen carrier gave a significant effect on the biomass conversion and improved the synthesis gas yield obviously. It was observed that the gas content of CO and H_2 was over 70% in CLG of biomass. The reduced hematite particles mainly existed in form of FeO. It was showed that the reduction of natural hematite with biomass proceeds in a stepwise manner from $Fe_2O_3 \rightarrow Fe_3O_4 \rightarrow$ FeO. Reduction product of natural hematite can be restored the lattice oxygen by oxidation with air.

INTRODUCTION

Chemical-looping gasification (CLG) is a novel gasification technique that involves the use of oxygen carrier, which transfers oxygen from air to the fuel, which is partially oxidized into H_2 and CO, avoiding the direct contact between them. Therefore, the air to fuel ratio is kept low to prevent the fuel from becoming fully oxidized to CO_2 and H_2O. [1]. A basic chemical-looping gasification system has two reactors, one for air and one for fuel, as is illustrated in Figure 1. On the contrast of traditional gasification technologies, CLG has several potential benefits as follows [2]. Firstly, the recycling of oxygen carrier can provide the oxygen needed for gasification, thus, it can save the cost for making pure oxygen. Secondly, the oxidation reaction of the metal oxide is very exothermic; however, the reduction reactions are endothermic. So, the heat for the endothermic reduction reactions is given by the circulating solids coming from the air reactor at higher temperature. And the same time, the oxygen carrier also can catalyze tar cracking, which reduced the content of tar in biomass gasification [3-5].

There are some works studying chemical looping gasification of solid fuels. He et al. [6] investigated the CLG of biomass with natural hematite as oxygen carriers in a bubbling fluidized bed reactor. It was found that the gasification efficiency and carbon conversion reached up to 75.8% and 94%, respectively. Cao et al. [7] tested in circulating fluidized bed reactor different oxygen carriers, which can capture that the concentration of CO_2 enriched to 99%. B. Acharya et al. [8] put forward a set of system which produced hydrogen through biomass chemicallooping gasification. Efficiency of the system was as high as 87.49%, and the hydrogen concentration was found to be 71%, in addition, zero discharge of CO_2 can realize.

The objective of this investigation was to explore the possibility of using natural hematite (Fe_2O_3) as the oxygen carrier in CLG of biomass. In the present work, the thermodynamics of biomass chemical looping gasification was analyzed, and the same time, CLG of biomass with natural hematite as oxygen carrier was experimentally investigated in a lab-scale bubbling fluidized bed reactor using argon as fluidizing gas.

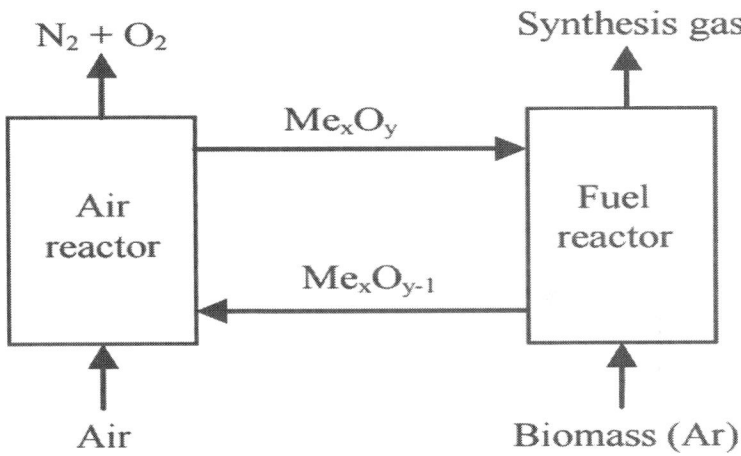

Figure 1: Chemical-looping gasification of biomass with oxygen carrier.

THERMODYNAMIC ANALYSIS OF CHEMICAL-LOOPING GASFICATION

Chemical Reactions

Fe_2O_3 as oxygen carriers in the air reactor, the main reactions at 750°C probably are:

$$4Fe_3O_4 + O_2 \rightarrow 6Fe_2O_3$$
$$\Delta H = -483.07 \text{ kJ/mol}$$

(1)

$$4FeO + O_2 \rightarrow 2Fe_2O_3$$
$$\Delta H = -554.68 \text{ kJ/mol}$$

(2)

In the fuel reactor, the main reactions probably are: pyrolysis of biomass:

$$C_nH_{2m}O_x \rightarrow char + tar + syngas$$
$$(CO, H_2, CO_2, CH_4)$$

(3)

The reduction reactions of oxygen carrier particle with pyrolytic products of biomass at 650°C [9]

$$CO + 3Fe_2O_3 \rightarrow CO_2 + 2Fe_3O_4$$
$$\Delta H = -37.67 \text{ kJ/mol}$$

(4)

$$CO + Fe_2O_3 \rightarrow CO_2 + 2FeO$$
$$\Delta H = -3.22 \text{ kJ/mol}$$

(5)

$$H_2 + 3Fe_2O_3 \rightarrow H_2O + 2Fe_3O_4$$
$$\Delta H = -2.01 \text{ kJ/mol}$$

(6)

$$H_2 + Fe_2O_3 \rightarrow H_2O + 2FeO$$
$$\Delta H = 32.44 \text{ kJ/mol}$$

(7)

$$CH_4 + 3Fe_2O_3 \rightarrow 2H_2 + CO + 2Fe_3O_4$$
$$\Delta H = 221.76 \text{ kJ/mol}$$

(8)

$$CH_4 + 3Fe_2O_3 \rightarrow 2H_2 + CO + 2Fe_3O_4$$
$$\Delta H = 221.76 \text{ kJ/mol}$$

(9)

$$CH_4 + 4Fe_2O_3 \rightarrow 2H_2O + CO_2 + 8FeO$$
$$\Delta H = 317.87 \text{ kJ/mol}$$

(10)

$$C + 3Fe_2O_3 \rightarrow CO + 2Fe_3O_4$$
$$\Delta H = 133.65 \text{ kJ/mol}$$

(11)

Therefore, in the presence of oxygen carriers, these reactions occur sequentially and simultaneously during biomass pyrolysis and gasification. The final products of biomass gasification are determined by the interaction of a couple of above mentioned reactions.

Thermodynamic Analysis of Gasification Process

Effect of the temperature on the Gibbs free energy (ΔrG) and chemical equilibrium constant (Logk) of reactions (4)-(11) was shown in Figure

2. In Figure 2, it found that the ΔrG is less than zero and the Logk is more than zero, which means that the reactions (4)-(11) can react in the thermodynamics. If dividing the reactions (4)-(11) into four groups, that is (4)-(5); (6)-(7); (8)-(9); (10)-(11), it can be seen that the former is less than the later for ΔrG, but the trend of Logk is opposite from the Figure 2, further, the trends of change increased with the increase of temperature. As we known, if the smaller ΔrG and the greater Logk, the reaction will be easier to occur. So, it can conclude that oxygen carriers were reduced gradually in the atmosphere of biomass pyrolysis. It meant the oxygen carriers changed as follows: $Fe_2O_3 \rightarrow Fe_3O_4 \rightarrow FeO$.

Chemical Equilibrium of Biomass Gasification Process

There were a number of reactions in the CLG according to the theory analysis. However, some reactions were predominant and the other reactions were secondary in the actual process. If the secondary reactions were ignored, it can help us a better understanding of the CLG. According to the principle of Gibbs free energy minimization, equilibrium components of biomass with Fe_2O_3 were investigated through HSC Chemistry 5.1 software, which is a produced by Outokumpu company in Finland. The approximate formula of biomass is $CH_{1.34}O_{0.65}$, regardless the S and N.

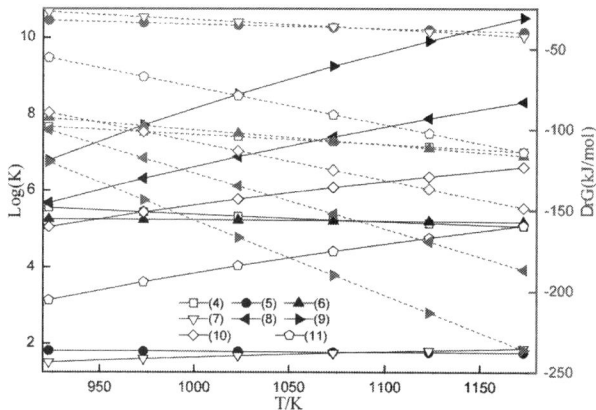

Figure 2: Effects of temperature on ΔrG and LogK of reaction (4)-(11).

In order to illustrate the influence of the biomass to Fe_2O_3 ratio on equilibrium composition, the equilibrium composition was calculated by changing Fe_2O_3 content at the same temperature. The result was shown in Figure 3. The temperature is set at 750°C and the initial value of $CH_{1.34}O_{0.65}$ is 1kmol in the process. In Figure 3(a), the Fe_2O_3 content was gradually increased from 0.02 kmol to 6 kmol. As seen in figure, the content of incomplete oxidation products $(CO + H_2)$ gradually decreased and the content of complete oxidation products $(CO_2 + H_2)$ gradually increased with the amount of Fe_2O_3 gradually increased, respectively. As the amount of Fe_2O_3 is nearly 5 kmol, the amount of $H_2 + CO$ is almost zero, and the content of $H_2O + CO_2$ is almost constantly. At the same time, the amount of carbon which produced by carbon deposit reaction decreased with increase the amount of Fe_2O_3, the carbon is almost disappeared as the Fe_2O_3 content near 1kmol. It can be observed the reduction products of Fe_2O_3 are mainly FeO and Fe_3O_4. As the amount of Fe_2O_3 is less than 3.5 kmol, the main reduction product is FeO, however, the main reduction product is Fe_3O_4 when the amount of Fe_2O_3 is more than 3.5 kmol. In this paper, it concerned to produce syngas which is mainly composed of H_2 and CO. Thus, in order to obtain syngas instead of CO_2 and water, it is necessary to keep the ratio of lattice oxygen to fuel as low as to prevent the fuel from being fully oxidized to CO_2 and H_2O. The effect of the amount of Fe_2O_3 from 0.02 to 1.2 kmol on the equilibrium composition is shown in detail in Figure 3(b). As shown in the figure, the amount of CO reaches maximum when the Fe_2O_3 content is about 0.2 kmol, and then which is declined gradually with the amount of Fe_2O_3 increasing. It is worth to note that the amount of CO_2 is more than CO as the Fe_2O_3 content is more than 0.7 kmol, which means that biomass is completely oxidized due to the lattice oxygen increasing. So, it can produce more CO_2 rather than CO. Therefore, the amount of Fe_2O_3 must be kept between 0.2 kmol and 0.7 kmol to obtain the syngas. The reduction product of Fe_2O_3 is mainly FeO, and almost nonexistent Fe_3O_4. At the same time, it can infer that carbon deposition was caused by material excessive from the Figure 3.

According to the analysis of the thermodynamic theories, it can obtain the synthesis gas which the main component is CO and H_2 through the ratio of fuel to lattice oxygen is kept low level. Meanwhile, the main reduction product of Fe_2O_3 is FeO.

In the next work, CLG of biomass with natural hematite as oxygen carrier was experimentally studied in a lab-scale bubbling fluidized bed reactor using argon as fluidized gas.

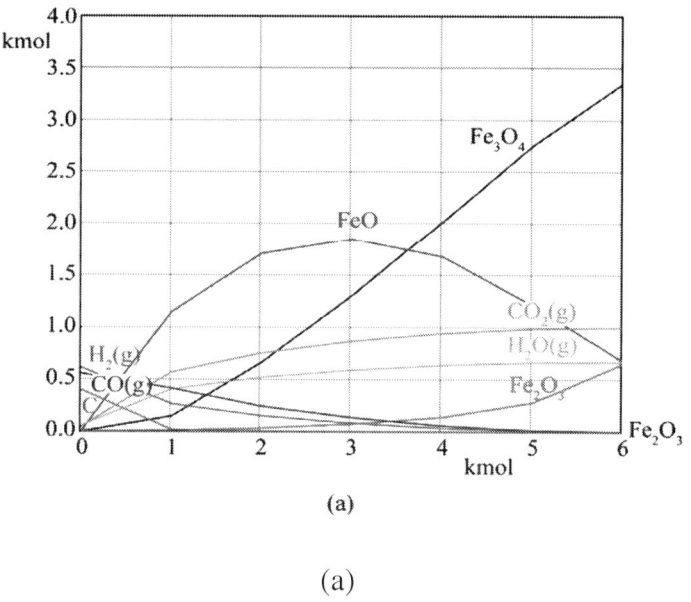

(a)

(a)

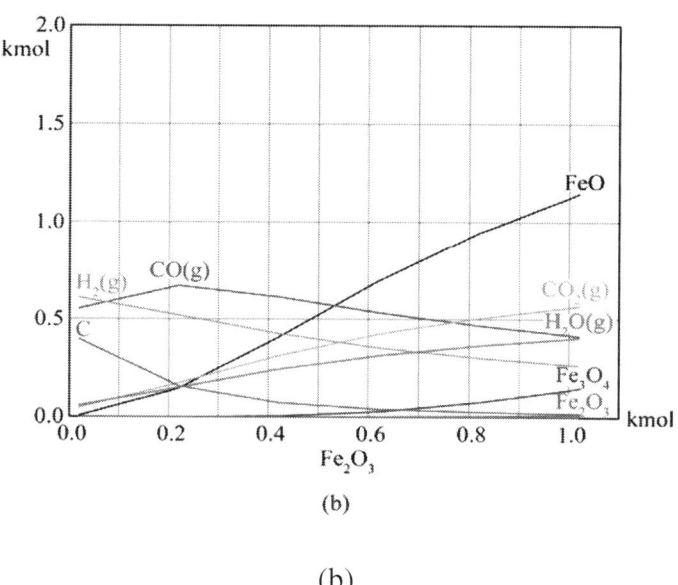

(b)

(b)

Figure 3: Effects of Fe_2O_3 on equilibrium composition at 750°C.

EXPERIMENTAL

Raw Materials

The sawdust of pine with particle sizes between 250 - 425 μm was dried in the oven which kept 105°C. The dry-basis proximate analysis and ultimate analysis of the sawdust are showed in Table 1. The hematite with particle sizes between 180 - 250 μm which was supplied by Guangdong Iron & Steel Group Co. Ltd. The elements composition analysis of the hematite is presented in Table 2. According to the Fe^{3+} fraction, it is calculated that the Fe_2O_3 content is about 81.66% in the hematite.

Laboratory Setup

The experiments were conducted with a bubbling fluidized bed reactor of quartz placed in a transparent furnace, as is illustrated in Figure 4. The reactor is in length of 1000 mm and an inner diameter of 60 mm. The bed temperature was measured 5 mm above the porous quartz plate using a K-type thermocouple. Oxygen carrier particles were placed on the porous plate and were then heated in air flow to set temperature. During the reducing period, the fluidizing gas was argon (800 L/hr), which was introduced from the bottom of the reactor. When the temperature comes to the set value, biomass was continuously fed (by a screw feeder) in a hopper at the top of the reactor. At the same time, argon (200 L/hr) was introduced from the top of the hopper. Therefore, the biomass sample was pushed by argon flow into the fluidized bed through a drop tube. The flue gases are passed through the cold trap to collect solid particles, water, and tar then introduced into the sampling bag. The composition of the gas products is measured using a gas chromatograph (SHIMADZUA Gas Chromatograph, GC-20B). All the oxygen carrier samples were performed with an X-Ray Diffraction (X'Pert PRO XRD).

Table 1: Proximate and ultimate analysis of pine

Proximate analysis wad %					Ultimate analysis wad %				QnKJ/ Kg
M	V	FC	A	C	H	O*	N	S	
8.39	84.31	6.88	0.42	49.66	5.55	43.33	0.021	1.44	18506

* : by differences

Table 2: Elements composition analysis of iron ore (%)

Item	S	Fe3+	Fe2+	SiO2	AlO3
Content	0.031	57.16	0.41	5.64	0.88
Item	CaO	MgO	Na2O	K2O	P
Content	o.24	0.035	0.10	0.048	0.017

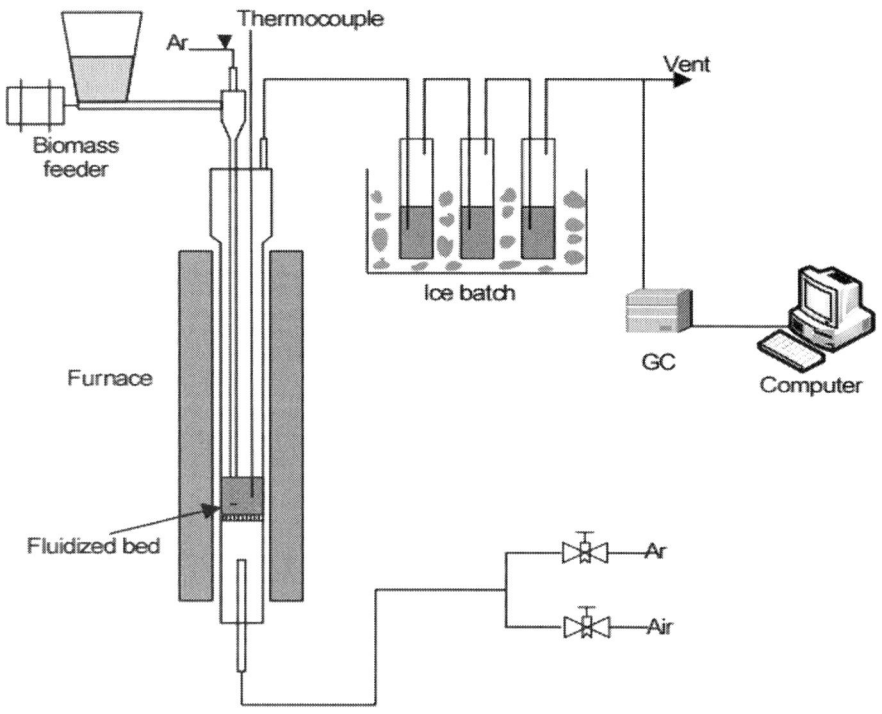

Figure 4: Schematic layout of the laboratory setup.

RESULTS AND DISCUSSION

Effect of the Presence of Hematite Particles on the Biomass Gasification

Based on the thermodynamic analysis, the mass of oxygen carriers and biomass material were set to 150 g and 44 g in a single test, respectively. The ratio of lattice oxygen to fuel is 0.42.

The effect of the presence of hematite particles on the CLG of biomass was investigated at temperatures of 750°C. Figure 5 shows the comparison of the gas yields of each gaseous component and the sum of the yields of H_2, CH_4, CO and CO_2 in the process of biomass CLG with those of blank tests during testing period of 30 min. In the blank tests, the hematite particles were replaced by silica sand. Therefore, the process taking place in the blank tests is only pyrolysis rather than CLG of biomass. As shown in Figure 5, the volume yields of each gaseous constituent were obviously enlarged in the presence of hematite particles. The total gas yields were increased from 25.17 L up to 42.39 L at 750°C. The yields of the generated gases (H_2, CH_4, CO, and CO_2) obtained with hematite particles were apparently larger than with silica sand, which suggest that the lattice oxygen in the hematite particles was used as the gasifying agent during biomass gasification. The main components of gaseous product are CO and H_2 which reached about 70% amount of the total volume of the gaseous product. The promoter action of lattice oxygen to carbon conversion was more obvious than to hydrogen conversion. So, the carbon can convert into carbon dioxide and carbon monoxide more completely, it can cause the volume of gaseous product which contained carbon element going up obviously, especially the volume of carbon dioxide. Figure 5 also illustrates gas concentrations (on the basis of argon free) of biomass gasification with oxygen carrier and pyrolysis with silica sand during 30 min at 750°C. Clearly, it was found that the concentrations of H_2, CH_4, CO and CO_2 were kept stable when biomass was pyrolyzed with inert silica sand. Whereas, there were significant changes for the gaseous concentration of biomass with hematite particles as reaction time increasing, especially the concentration of CO_2 and H_2. The CO and CH_4 concentrations of biomass with hematite particles were both lower than those of biomass with silica sand. H_2 concentration increased with the reaction time going on in the CLG of biomass probably owing to a part of biomass was totally oxidized into CO_2 and H_2O which subsequently accelerated the hydrogen generation reactions taking place rightward. In addition, in the initial stage of biomass gasification with hematite particles, the CO_2 fraction was more than 30% in the generated gas due to the easily availability of active lattice oxygen in the oxygen carrier particles. As the reactions proceeded, the concentration of CO_2 decreased gradually to 10% approximately. These results may be contributed to the amount of valid hematite particles decreasing with the reactions proceeding. So, portion of biomass taken out by fluidized gas after only pyrolytic reaction, which can cause the proportion of pyrolysis gas increasing in the gaseous products. Through the contrast test, it experimental verified the feasibility of biomass chemical-looping gasification with hematite particles as oxygen carriers.

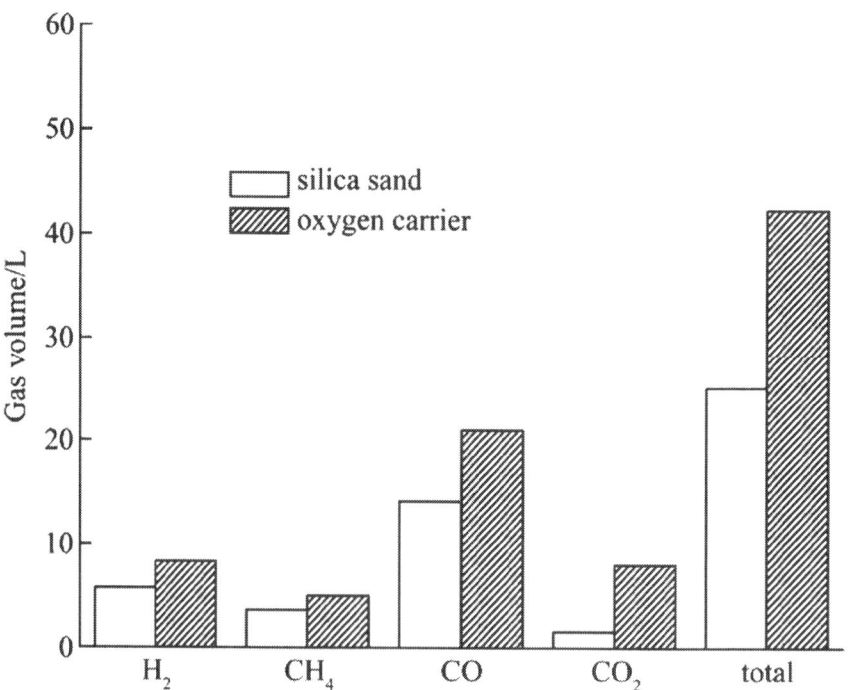

(a)

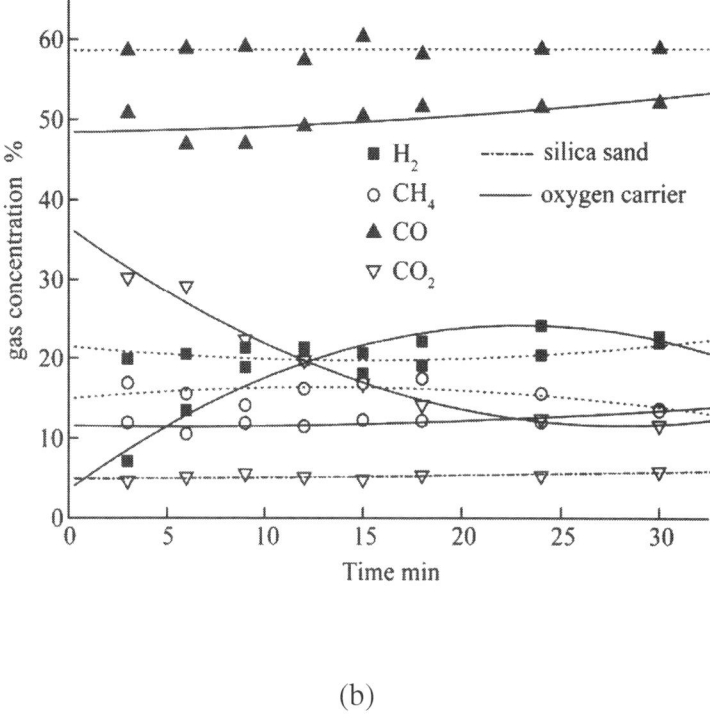

(b)

Figure 5: The effect of oxygen carrier on CLG of biomass at 750°C.

XRD Analysis of Oxygen Carriers

Figure 6 shows the XRD patterns of fresh oxygen carrier, the oxidized oxygen carrier with the air, and the reduced oxygen carrier of the fuel reactor under different temperature, respectively. InFigure 6(b), it can be seen that the mainly existed in form of Fe_2O_3 and SiO_2 for fresh oxygen carriers. Comparison to fresh oxygen carrier, there was no chemical changes occurred in the regeneration of reacted oxygen carrier, and Fe_2O_3 was found the major phase with minor amounts of Fe_3O_4. The forms of oxygen carriers after the reduction reaction changed as temperature rising, as seen in Figure 6(a), at lower temperature (650°C), the variation of oxygen carriers was from Fe_2O_3 to Fe_3O_4, the main existence form of reduced oxygen carriers was FeO at higher temperature (≥750°C).

However, the variation of hematite particles as oxygen carriers was from Fe_2O_3 to Fe_3O_4 in the CLC of biomass [10-12].

The elements content of Fe in hematite particles was analyzed as shown in Figure 7, it found that the ferric iron was reduced to ferrous iron in the CLG of biomass. The tendency of conversion was more and more obviously with increase of temperature. The capacity of lattice oxygen in oxygen carriers was recovered by air oxidization, which was affected slightly on the reduction temperature. So, it is feasible that the hematite particles can be used circularly as oxygen carriers in the CLG of biomass, which have higher oxygen transfer ability.

According to the analysis of existence form of reduced oxygen carriers, it found that hematite particles undergone the process of biomass gasification as oxygen carriers, and the mainly existence form of hematite particles were FeO after reduction as the temperature higher than 750°C. Further, the reduction of hematite particles with biomass proceeded in a stepwise manner from Fe_2O_3 Fe_3O_4 FeO. So, the main reactions were (2), (5), (7), (9), and (11) in the CLG of biomass with the hematite particles as oxygen carriers.

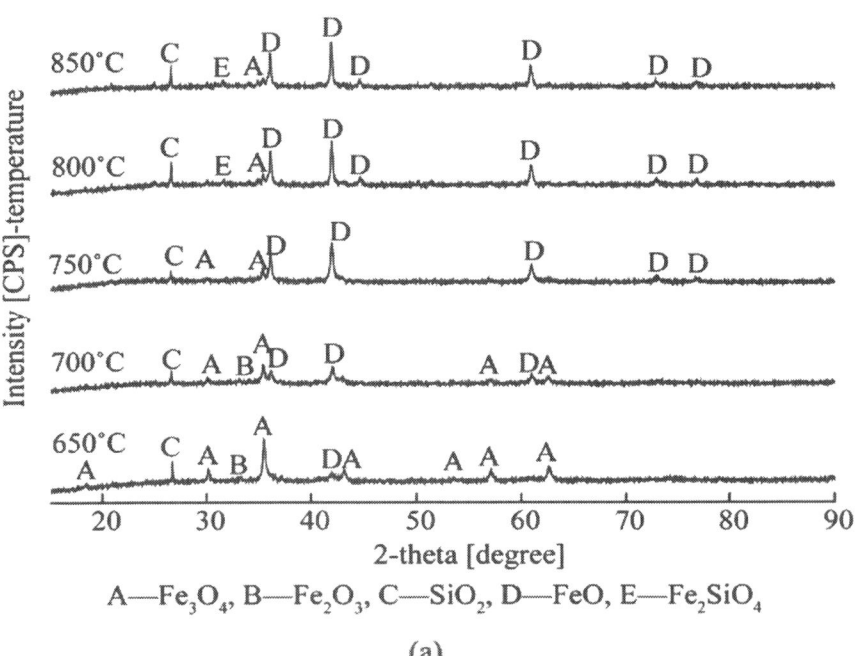

A—Fe_3O_4, B—Fe_2O_3, C—SiO_2, D—FeO, E—Fe_2SiO_4

(a)

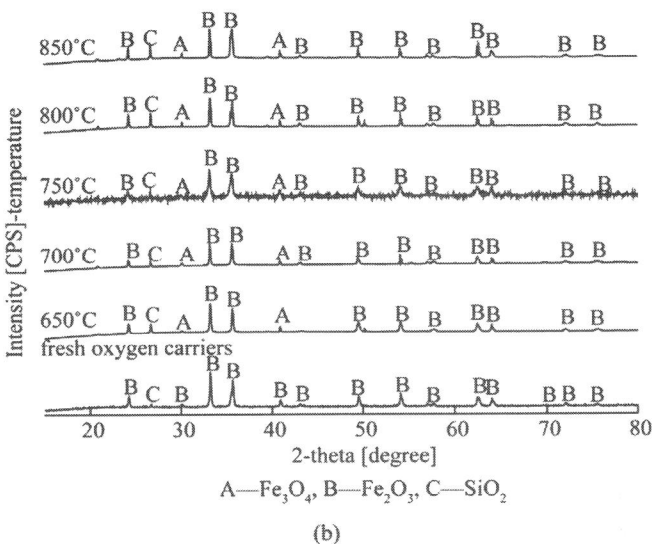

(b)

Figure 6: XRD patterns of fresh and reacted hematite particles at different temperatures. (a) Reduction; (b) Oxidization.

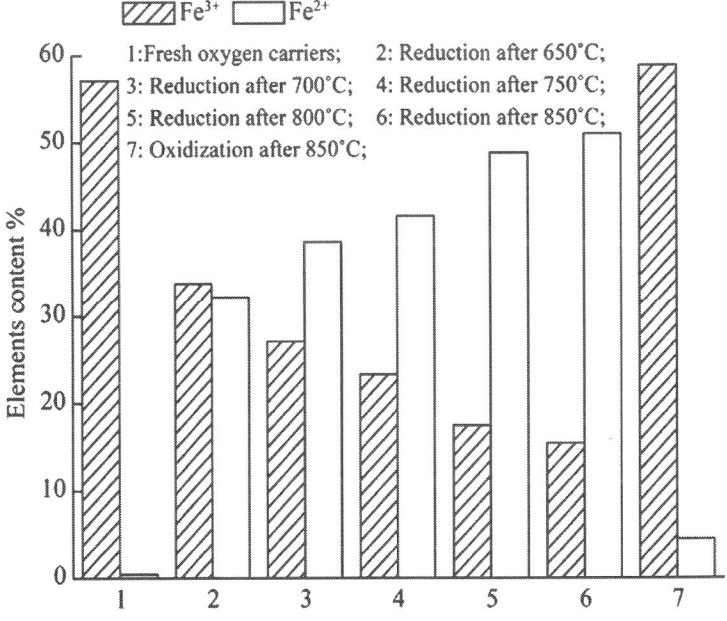

Figure 7: Elements composition analysis of hematite particles.

CONCLUSIONS

The process in CLG of biomass mixed with natural hematite as oxygen carriers was analyzed using the theory of thermodynamics, and then, the verification test was studied in a bubbling fluidized bed reactor with argon as fluidizing gas. It was found that the theory analysis was coincidence with the results of experiment. Consequently, it is feasibility natural hematite as oxygen carriers in CLG of biomass. It can obtain synthesis gas which mainly included CO and H_2 by limiting the ratio of lattice oxygen to fuel between 0.2 and 0.7. The presence of oxygen carriers could obviously affect the process of biomass thermal conversion, which can significantly increase gas yield and carbon conversion rate. The main components of gaseous product were CO and H_2 which reached about 70% amount of the total volume of the gaseous product. The concentration of CO was the highest and the concentration of CH_4 was the lowest during the biomass gasification. In addition, the concentrations of H_2, CO, and CH_4 in the product gas slowly increased with the reaction proceeding, and the CO_2 concentration showed an opposite trend. XRD analysis showed that the iron element in the reduced hematite particles mainly existed in form of FeO, with minor formation of Fe_3O_4. Further, the lattice oxygen released, corresponding to the transformation $Fe_2O_3 \rightarrow Fe_3O_4 \rightarrow FeO$, provided the oxygen element needed in biomass gasification. The capacity of lattice oxygen in reduced oxygen carriers can be recovered through air oxidization.

ACKNOWLEDGEMENTS

The financial support of National Natural Science Foundation of China (51076154) is gratefully acknowledged. This work was also supported by Science and Technology Project of Guangdong (2010B010900047), "12th Five Years" National Science and Technology Support Program (2011BAD15B05).

REFERENCES

1.   J. Bolhar-Nordenkampf, T. Proll, P. Kolbitsch, et al., "Performance of a NiO-Based Oxygen Carrier for Chemical Looping Combustion and Reforming in a 120 kW Unit," Energy Procedia, Vol. 1, No. 1, 2009, pp. 19-25. doi:10.1016/j.egypro.2009.01.005

2.   L. F. De Diego, M. Ortiz, F. Garcia-Labiano, et al., "Synthesis Gas Generation by Chemical-Looping Reforming Using a Nibased Oxygen Carrier," Energy Procedia, Vol. 1, No. 1, 2009, pp. 3-10. doi:10.1016/j.egypro.2009.01.003

3.   K. Tomishige, T. Kimura, J. Nishikawa, et al., "Promoting Effect of the Interaction between Ni and CeO_2 on Steam Gasification of Biomass," Catalysis Communications, Vol. 8, No. 2, 2007, pp. 161-166.

4.   T. Nordgreen, T. Liliedahl and K. Sjostrom, "Metallic Iron as a Tar Breakdown Catalyst Related to Atmospheric, Fluidized Bed Gasification of Biomass," Fuel, Vol. 85, No. 5-6, 2006, pp. 689-694. doi:10.1016/j.fuel.2005.08.026

5.   C. Courson, E. Makaga, C. Petit, et al., "Development of Ni Catalysts for Gas Production from Biomass Gasification Reactivity in Steam and Dry Reforming," Catalysis Today, Vol. 63, No. 2, 2000, pp. 427-437. doi:10.1016/S0920-5861(00)00488-0

6.   F. He, Z. Huang, H. B. Li, et al., "Biomass Direct Chemical Looping Conversion in a Fluidized Bed Reactor with Natural Hematite as an Oxygen Carrier," Proceedings of the 2011 Asia-Pacific Power and Energy Engineering Conference, Wuhan, 25-28 March 2011.

7.   Y. Cao and W.-P. Pan, "Investigation of Chemical Looping Combustion by Solid Fuels. 1. Process Analysis," Energy & Fuels, Vol. 20, No. 5, 2006, pp. 1836-1844. doi:10.1021/ef050228d

8.   B. Acharya, A. Dutta and P. Basu, "Chemical-Looping Gasification of Biomass for Hydrogen-Enriched Gas Production with In-Process Carbon Dioxide Capture," Energy Fuel, Vol. 23, No. 10, 2009, pp. 5077-5083.

9.   A. Abad, J. Adanez, F. Garcia-Labiano, et al., "Mapping of the Range of Operational Conditions for Cu-, Fe-, and Ni-Based Oxygen Carriers in Chemical-Looping Combustion," Chemical

Engineering Science, Vol. 62, No. 1-2, 2007, pp. 533-549. doi:10.1016/j.ces.2006.09.019

10. L. H. Shen, J. H. Wu, J. Xiao, et al., "Chemical-Looping Combustion of Biomass in a 10 kW_{th} React or with Iron Oxide as an Oxygen Carrier," Energy and Fuels, Vol. 23, No. 5, 2009, pp. 2498-2505. doi:10.1021/ef900033n

11. H. Leion, E. Jerndal, B. M. Steenari, et al., "Solid Fuels in Chemical-Looping Combustion Using Oxide Scale and Unprocessed Iron Ore as Oxygen Carriers," Fuel, Vol. 88, No. 10, 2009, pp. 1945-1954. doi:10.1016/j.fuel.2009.03.033

12. A. Rubel, K. L. Liu, J. Neathery, et al., "Oxygen Carriers for Chemical Looping Combustion of Solid Fuels," Fuel, Vol. 88, No. 5, 2009, pp. 876-884.doi:10.1016/j.fuel.2008.11.006

Methodology of Supervision by Analysis of Thermal Flux for Thermal Conduction of a Batch Chemical Reactor Equipped with a Monofluid Heating/Cooling System

Ghania Henini[1, 2], Fatiha Souahi[2],
and Ykhlef Laidani[1]

[1]Hassiba Ben Bouali University, Hay Essalem, Chlef, 02000, Algeria
[2]Ecole Nationale Polytechnique, El Harrach 16200, Algeria

ABSTRACT

We present the thermal behavior of a batch reactor to jacket equipped with a monofluid heating/cooling system. Heating and cooling are provided respectively by an electrical resistance and two plate heat exchangers. The control of the temperature of the reaction is based on the supervision system. This strategy of management of the thermal devices is based on the usage of the thermal flux as manipulated variable. The modulation of the monofluid temperature by acting on the heating power or on the opening degrees of an air-to-open valve that delivers the monofluid to heat exchanger. The study shows that the application of this method for the conduct of the pilot reactor gives good results in simulation and that taking into account the dynamics of the various apparatuses greatly improves ride quality of conduct. In addition thermal control of an exothermic reaction (mononitration) shows that the consideration of heat generated in the model representation improve the results by elimination any overshooting of the set-point temperature.

INTRODUCTION

A large number of industrial processes, such as the production of polymers, fine chemicals, and pharmaceuticals for which continuous production is not feasible or economically attractive, are operated in batch reactor.

In many cases this mode of operation is used to manufacture a variety of products that involve significantly different characteristics such as the conversion time and heat of the reaction. The control of such a type of reactors is completely often difficult to achieve [1, 2] due to their flexible and multipurpose utilization (different operating configurations and use for different productions). To guarantee batch-to-batch reproducibility and improve yield and selectivity, automation of batch reactors must be largely increased. Due to the complexity of the reactions mixture and the difficulties to perform on-line composition a measurement, control of batch and fed-batch reactors, is essentially a problem of temperature control [3, 4]. Batch and fed-batch reactors require good temperature control due to the existence of heat-sensitive chemical reactants and/or products and also to the dependency of reaction rate on the temperature. To carry out chemical reactions in this type of reactors, frequently an operating mode consisting of the different phases is used:

- A heating phase of the reaction mixture until the desired reaction temperature,

- A reaction phase during which the temperature is maintained constant,
- A cooling phase to avoid by-products formation [5–7].

Any controller was used to controlling the reactor must be able to take into account these different stages [8–11]. The temperature of the reactor content is controlled by the heat exchange with a fluid flowing inside the jacket surrounding the reactor. It is complicated by the operating mode and the numerous fluids to be managed. The control performances are then mainly dependent on the heating-cooling system associated with the reactor [12]. Many studies have been performed on control problems and strategies in such a type of reactors [13, 14].

Many configurations of heating/cooling systems are cited in the literature [13]. Two main types of heating/cooling systems are commonly used in industry: the alternative system or multifluid and mono-fluid system [15, 16]. With the multifluid system, the utility fluids flow alternatively in the jacket depending on the required control variable. The utility fluids are available to a given temperature (plant utilities). This system represents the most largely used system in industry due to the relative ease of design and low cost (direct use of plant utilities). Therefore, the control task of this type of process is rather difficult and can be divided in two parts: firstly, the choice of the right fluid and, secondly, the appropriate action on the flow-rate of this utility fluid in order to satisfactorily track the desired temperature profile. Thus, to go from heating to cooling, a change of fluid is required which results in a discontinuity in the operation.

An alternative configuration is the mono-fluid system which tends to replace the multifluid system in recently built workshops. This system uses only one fluid, the temperature of which can be modified to achieve the desired reactor temperature by an intermediate thermal loop, which may include heat exchangers, power heaters, and so forth; nevertheless, the dynamics of this external thermal loop can be penalizing, particularly in the case of urgent need of a rapid cooling or heating [17]. This technique has been studied earlier [15, 18].

The principle of functioning consists in choosing the apparatus or the fluid to be injected into the thermal loop and then to determine the manipulated variable, which will make it possible to follow a given profile of temperature. This variable can be an electrical power or a flow-rate. This choice is carried out by a program of supervision. Many authors have studied both heating/cooling systems and the supervision. Among the authors [13] presented experimental and simulation studies for temperature control of industrial batch reactors (16, 65, and 160l) equipped with a multifluid heating/cooling systems. They used adaptive GPC with double models references and the supervision strategy based one the limits thermal flux analysis. A new tuning approach of the GPC algorithm has been proposed by [19] in both adaptive and non-adaptive configurations for a fed-batch penicillin process using the complete factorial design method. Another approach for the management of the fluids was developed by [20]. This approach is based on the principle of the supervision by prevision of limit trajectories of temperature. The tool of supervision calculates two criteria every period of sampling, with regard to the distance enter the set-point temperature of the reaction mixture and the evolution which would present the system with a maximum and minimum thermal flux of each fluid.

An effective approach to the control of reactor and similar processes utilizes various methods of the nonlinear control (NC). Several modifications of the NC theory are described in, for example, [21, 22]. Especially, a large class of the NC methods exploits linearization of nonlinear plants, for example, [23], an application of PID controllers, for example, [24], or factorization of nonlinear models of the plants on linear and nonlinear parts, for example, [25–27].

This paper deals with a new methodology of supervision and control for the thermal behavior of reactor equipped with a mono-fluid heating/cooling system.

A master controller calculates the thermal flux (can be exchanged between the reaction and the jacket) necessary to flow the required reactor temperature profile. On the other hand, the maximum and minimum thermal capacities of each thermal fluid are determined and used to choose the "right" fluid with the priority to the fluid present in the jacket. The required thermal flux calculated by the master controller is compared to the limit capacities of the present fluid. The master controller is a Predictive Functional Control (PFC). Chosen for its capability to maintain the flexible character of reactor. After a description of the reactor (pilot plant). In the third part, methodology for monitoring and the control of batch reactors are described. Then a brief description of the procedure of regulation (PFC). One gives the strategy of supervision in the fourth part. The last part is devoted to the validation of the system of supervision based on the strategy of management of the thermal devices by the analysis of heat flux. Obtained results, for the thermal behavior of batch reactor in simulation following profiles of typical temperatures (heating-maintenance-cooling-maintenance).

PROCESS DESCRIPTION

A schematic diagram of the pilot plant is depicted in Figure 1.

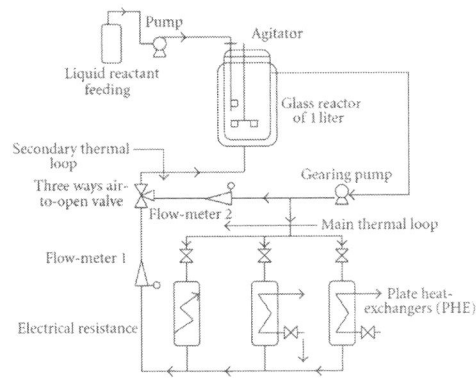

Figure 1: Schematic of the pilot-plant.

The experimental device consists of a 1 L jacketed glass reactor, equipped with a mono-fluid heating/cooling system. The mono-fluid used in this work is a mixture of ethylene glycol and water, in a ratio of 50% in weight, with a flow rate of 1000 L h^{-1} and at a temperature which varies between −35 and 110°C. the mono-fluid flow-rate is measured by means of two flow-meters, one installed on the main thermal loop (flow-meter 1) and the other one on the secondary thermal loop (flow-meter 2). The heating/cooling system uses an electrical resistance of 2000 W and two plate heat-exchangers (PHE). One PHE uses cold water as a utility fluid at a temperature around 15°C and a maximum flow rate of 1500 L h^{-1} while the other one uses a mixture of ethylene glycol and water, in a ratio of 50% in weight, at a temperature around of −10°C and a maximum flow of 1500 L h^{-1}. Flow rates of the utility fluids are also measured. Three on-off valves allow the mono-fluid to be heated or cooled. Two other on-off valves are used to manipulate the utility fluids. A three-way air-to-open valve ensures the division of the mono-fluid in two parts during the cooling phases. A gearing pump ensures the circulation of the mono-fluid in the thermal loop at maximum flow rate of 1500 L h^{-1}. The reactor has the following physical specifications: an internal diameter of 82 mm, a wall thickness of 9 mm, an external jacket diameter of 125 mm, a jacket wall thickness of 5 mm, a maximal reactant mixture-reactor heat transfer area of 0.039 m^2 and a jacket volume of 0.15 L. A propeller rotated at 260 rpm. The reactor is operated in batch and fed-batch modes. A piston pump allows the variation of the liquid reactant flow rate from 0 to 336 cm^3 h^{-1}.

All the temperatures are measured at each sampling period using PT100 platinum resistance sensors with a precision of ±0.1°C. The feed temperature of inlet reactant is measured by a thermocouple. A computer with analog-to-digital (A/D) and digital-to-analog (D/A) converters is employed for data acquisition. The supervision and control algorithms programs are written in Fortran 4 and are implemented on a PC in order to accomplish the different temperature control tasks.

METHODOLOGY FOR MONITORING OF BATCH REACTOR

A strategy integrating the followup and control was developed by [13, 28, 29]. For a batch reactor of 16 L equipped with a heating/cooling system grouping together at once of multifluid and the mono-fluid. To realize the thermal behavior of this process, it also used generalized adaptive predictive control with double model reference (GPCMR).

For the study of the behavior of the process, the structure of regulation in cascade using the technique of predictive functional control (PFC) is application year. Two control levels "1" and "0" are considered. Two controllers are used in the control level "1." The first one in charge of supervision (PFCM), it computes the required thermal flux to be exchanged between the mono-fluid flowing inside the jacket and the reaction mixture. The second one (PFC1) is devoted to calculate the set-point temperature (T_{icons}) which will be tracked by the inlet jacket temperature. In the control level "0," because the algorithm of the PFC is an algorithm SISO (single input, single output), three slave controllers are used to control the mono-fluid temperature. One placed on the electrical resistance which computes the electrical power value and the other two on the two plant heat exchangers to compute the mono-fluid flow rate fraction dispatched to one of them.

The plan block for cascade control of this process is given in Figure 2.

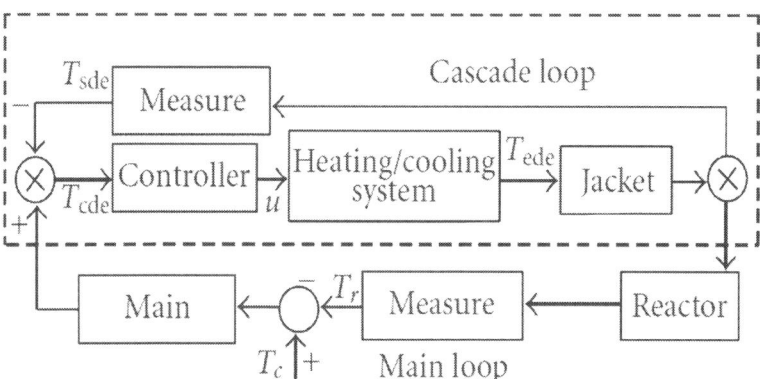

Figure 2: Schematic of the plant block en cascade control.

The model has been established using the following basic assumptions:

For the reactor.

- Specific heat and density of pure components are constant in the considered range of temperature, the heats of mixing are neglected.
- The reactions are carried out in a pseudohomogeneous liquid phase.
- If the reactions would be put into play, the heats of reactions are taken into account.
- The reactor is perfectly mixed with homogeneous temperature and concentrations in the reaction mixture.
- Feeding of the reactive is performed without volume concentration.
- The kinetic laws of the chemical reactions involved are assumed to be such that the reaction rate constant follows the Arrhenius law and the reaction rate is a function of the reactant concentrations.

Electrical Resistance

To model the electrical resistance, we considered spaced out annular as a block stirred, in which the monofluid exchanges some heat with the wall of the stem and receives a heat flux resulting from the electric resistance.

Heat Exchangers

Heat and cold currents are always parallel between the input and output device.

No the phase change and the physical properties of both fluids and plate remain constant in the temperature range used.

Heat loss with the external environment is considered negligible.

The flow of the two fluids is of piston type. For each fluid, the speed is the same on any plane normal to the direction of the currents, so there is only temperature gradient in the direction parallel to the currents.

The total exchange surface is the sum of the exchange surfaces of all plates.

The writing the thermal balances, reflecting the exchange between the reaction mixture and the jacket of the reactor, led us to use the thermal flux as a control variable intermediary. This variable can then simultaneously solve the problem of supervision and control. Moreover, this control strategy is applicable to any thermal system.

Thermal Behavior of the Reactor

The reactor can be described by the following thermal balances.

Thermal balance on the reaction mixture:

$$\rho_r C p_r V_r \frac{dT_r}{dt} = U A_{r,\mathrm{pr}}(T_{\mathrm{sde}} - T_r) + F_c \rho_c C p_c (T_c - T_r) - Q_r. \tag{1}$$

Initial conditions: the physico-chemical properties (ρ, Cp, V, U, F, Q, T_c) are data (constants).

T_{sde} calculated by (11).

Thermal balance on the mono-fluid flowing inside the jacket:

$$\rho_{fc} C_{p fc} V_{de} \frac{dT_{sde}}{dt} = UA_{r,pr}(T_r - T_{sde}) + F_{fc}\rho_{fc}C_{p fc}$$

$$\times (T_{ede} - T_{sde}).$$

$$(2)$$

Initial conditions: ρ_{fc}, F_c, $C_{p fc}$, V_{de}, $UA_{r,pr}$, F_{fc} are data. T_{ede} is the temperature of the utility used by the cooling heat exchanger: $T_e = T_{fc0}$ (cold water), $T_e = T_{fc0}$ (glycol water).

Tow controllers are used in the main controller loop, the first one called PFCM in charge of supervision; and the second one called PFC1 devoted to compute the jacket set-point temperature used by the slave controllers.

The internal model used by the PFCM controller was derived from (1) and it is given by the following continuous system transfer function equation using the Laplace transform:

$$T_r(p) = \left[\frac{1}{UA_{r,pr}} \right] \frac{1}{\tau_r p} \left(U_q(p) + Q_{ech}(p) \right)$$

$$T_r(p) = \frac{K}{p} \left(U_q(p) + Q_{ech}(p) \right),$$

$$(3)$$

where $\tau_r = \rho_r C_{pr} V_r / UA_{r,pr}$ time constant of the reactor, $U_q = UA_{r,pr}(T_{sde} - T_r)$ the thermal flux exchanged between reaction mixture and the mono-fluid present inside the jacket, $Q_{ech} = F_{cpc} C_{pc}(T_c - T_r) - Q_r$ the disturbance caused by feed reactants and the heat released by the chemical reaction, $K = 1/\tau_r UA_{r,pr}$ is the steady-state reactor gain (PFCM controller).

The following model was proposed to construct the PFC1 controller:

$$T_r(p) = \frac{K_r}{1 + \tau_r p} T_{ede}(p) + \frac{K_c}{1 + \tau_r p} T_c(p) + \frac{K_{Qr}}{1 + \tau_r p} Q_r(p).$$

$$(4)$$

where K_r is the steady-state reactor gain (PFC1 controller), K_c the steady-state liquid reactant feed gain, K_{Qr} the steady-state heat released gain.

SUPERVISION STRATEGY

The objective of the thermal behavior of the batch or fed-batch reactors equipped with a mono-fluid heating/cooling system is to be simultaneously able to determine the element of the thermal loop which is going to modify the temperature of the mono-fluid, as well as commands it to apply to the chosen actuator, to reach the objective of the regulation. The protocol of the thermal conduct is based on the following concept.

The regulator calculates the control variable which is interpreted as a thermal flux. This last one is then compared with thermal flux limits, calculated on-line for the various devices to choose the adequate element of the thermal loop. Once this choice realized, the manipulated variable is calculated according to the chosen device.

Limit Thermal Capacities

Electrical Resistance

This element of the thermal loop is used to heat the mono-fluid that can receive a maximum electric power $Pelec$ (max).

This maximum power will be considered as the maximum flux that can be exchange the mono-fluid with the reaction mixture through the jacket.

The evolution of the outlet temperature of the electrical resistance is given by the following thermal balance:

$$\left(\frac{V_{ch}}{F_{fc}}\right)\frac{d\left(T_{ch}-T_{ch}^{E}\right)}{dt} = \left(T_{ch}-T_{ch}^{E}\right) + \frac{P_{elec}(\max)}{\rho_{ch}Cp_{ch}V_{ch}}U_{p},$$

$$0 \le U_{p} \le 1. \qquad (5)$$

We distinguish two limits maximum thermal flux (Q_{maxch}) and minimum (Q_{minch}), such as

$$Q_{\text{max ch}} = P_{\text{elec}}(\text{max})$$

$$Q_{\text{min ch}} = 0.0. \tag{6}$$

Plate Heat Exchangers

It uses two thermal balances to calculate the thermal flux limit corresponding to these two heat exchangers. The first concerns the mono-fluid during its passage through the jacket and the other is the interchange between the reaction mixture and mono-fluid present in the jacket.

These balance sheets are translated by following equations:

$$Q_1 = F_{\text{fc}} \times \rho_{\text{fc}} \times Cp_{\text{fc}}(T_e - T_s), \tag{7}$$

$$Q_2 = U \times A_{r,\text{pr}}\left(\frac{(T_e - T_r) + (T_s - T_r)}{2}\right), \tag{8}$$

$$\frac{1}{U} = \left[\frac{1}{U_{r,\text{pr}}} + \frac{1}{U_{\text{pr,fc}}} + \frac{A_{r,\text{pr}}}{A_{\text{pr,fc}}}\right]. \tag{9}$$

The whole methodology, for the computation of the control variable, relies on the assumption that both thermal fluxes are equal:

$$Q_1 = Q_2. \tag{10}$$

From (7), (8) and (10), the outlet temperature of the jacket is calculated by the following relation:

$$T_s = \frac{\left(F_{\text{fc}} \times \rho_{\text{fc}} \times Cp_{\text{fc}} - U \times A_{r,\text{pr}}/2\right) \times T_e + U \times A_{r,\text{pr}} \times T_r}{F_{\text{fc}} \times \rho_{\text{fc}} \times Cp_{\text{fc}} + U \times A_{r,\text{pr}}/2}. \tag{11}$$

The flow rate of the mono-fluid being constant, during the cooling the minimal temperature which can be reached the entry of the jacket, is the one of the temperature of the utility of cooling used by the heat exchanger. That makes it possible to write:

$T_e = T_{fc0}$ (Cold water): in the case of the heat exchanger using cold water.

$T_e = T_{fc0}$ (Glycol water): in the case of the heat exchanger using the glycol water.

To calculate these minimal thermal flux Q_{minef} and Q_{mineg}, corresponds to the use of heat exchangers with cold water and with glycol water respectively, we will use the equations (7) or (8) by taking as external temperature of T_s that calculated by the equation (11).

We consider that the maximum thermal flux, corresponding to the use of heat exchangers with cold water and with glycol water respectively, is zero.

When a thermal equilibrium at the level of the double envelope, $(T_e = T_s)$ will occur:

$$Q_{max\ ef} = 0,$$

$$Q_{max\ eg} = 0. \tag{12}$$

Supervision Algorithm

Figure 3 gives an approximate global representation of the evolution of the thermal flux capacities of the different fluids as a function of the reactor temperature; the minimum and maximum capacities correspond to the border of the zones. Three zones are distinguished.

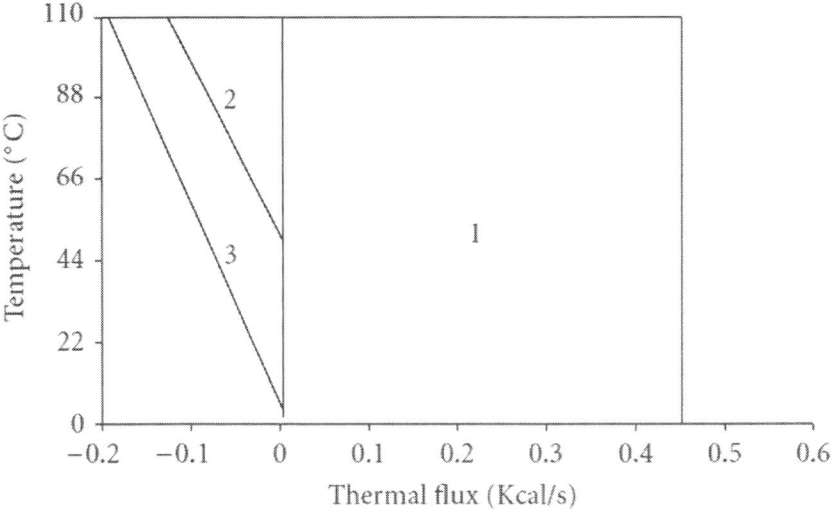

(1) Electrical resistance
(2) Plate heat-exchanger (cold water)
(3) Plate heat-exchanger (glycol water)

Figure 3: Evolution of the thermal flux capacities for the different fluids.

Zone 1: use the electrical resistance.

Zones 2: use of the heat exchanger with cold water.

Zones 3: use of the heat exchanger with glycol water.

When the thermal flux takes a positive value, it corresponds to a demand for heating and only the electrical resistance is concerned (phase = 1). A negative value means a request of cold; in this case one of both heat exchangers will be used according to the value of the thermal flux. So, a value in the range (Q_{minef}, Q_{maxef}), means a request of cooling by cold water, then only the heat exchanger in cold water is concerned (phase = 2) a value in the range (Q_{mineg}, Q_{minef}), means a request of cooling by the glycol water and only the heat exchanger for glycol water is concerned, (phase = 3). A flow chart of the overall procedure is given in Figure 4.

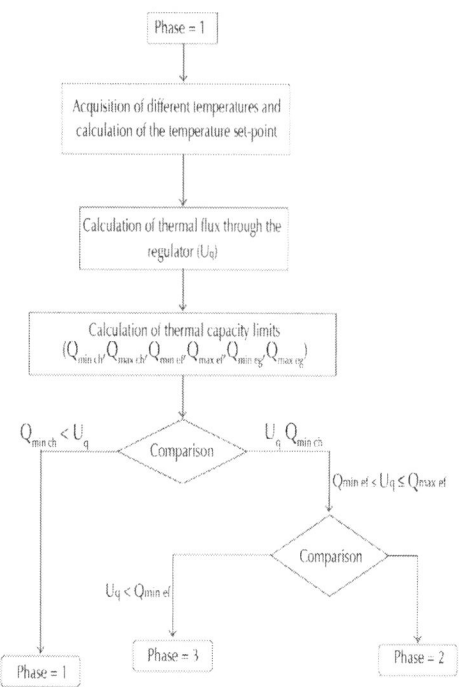

Figure 4: Flow chart of the supervisory and control in the case of the electrical resistance.

RESULTS AND DISCUSSION

In this paper, we present the simulation results obtained for the temperature control of an exothermic chemical reaction of acid-base neutralization between the hydrochloric acid (HCl) and the sodium hydroxide (NaOH).

The set-point temperature profile is composed by first stage: heating from 20 to 40°C during 1000 s; second stage: constant temperature at 40°C during 1500 s; third stage: cooling from 40 to 15°C during 1200 s and fourth stage: maintain at 15°C during 500 s.

However, the temperature profile is composed by an increase ramp-maintain and decrease ramp-maintain. In order to eliminate discontinuities resulting from change of stage, the set-point profile was filtered by a procedure called "docking procedure" [8, 17, 30]. The results are presented in Figure (5–7).

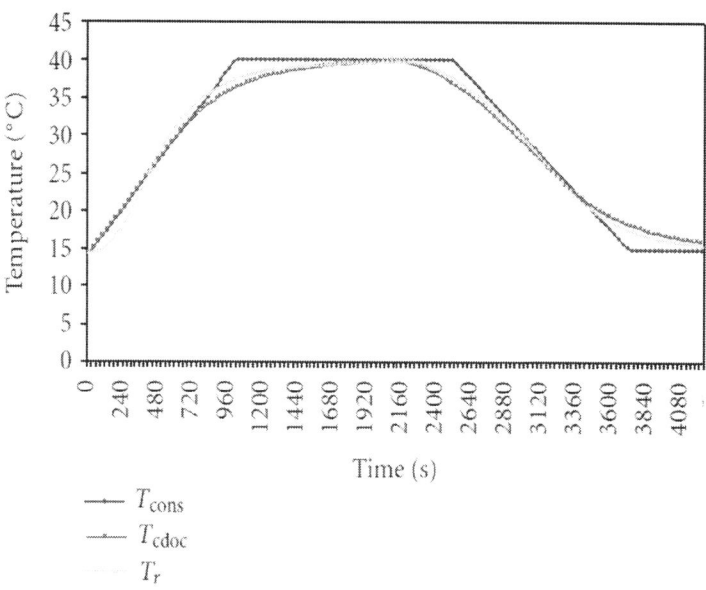

Figure 5: Evolution of temperature (T_{cons}), the adjusted set-point (T_{cdoc}), and reaction mixture (T_r), K = 0.5, Kr = 1.

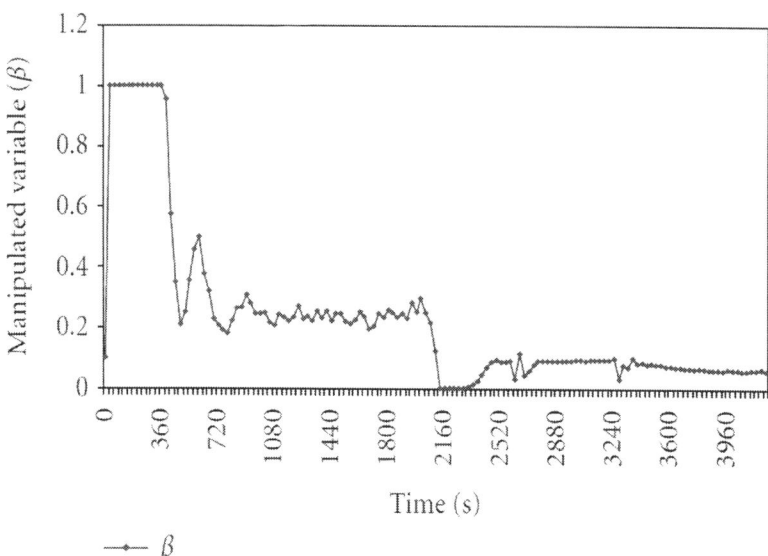

Figure 6: Evolution of the manipulated variable (β), K = 0.5, K_r = 1.

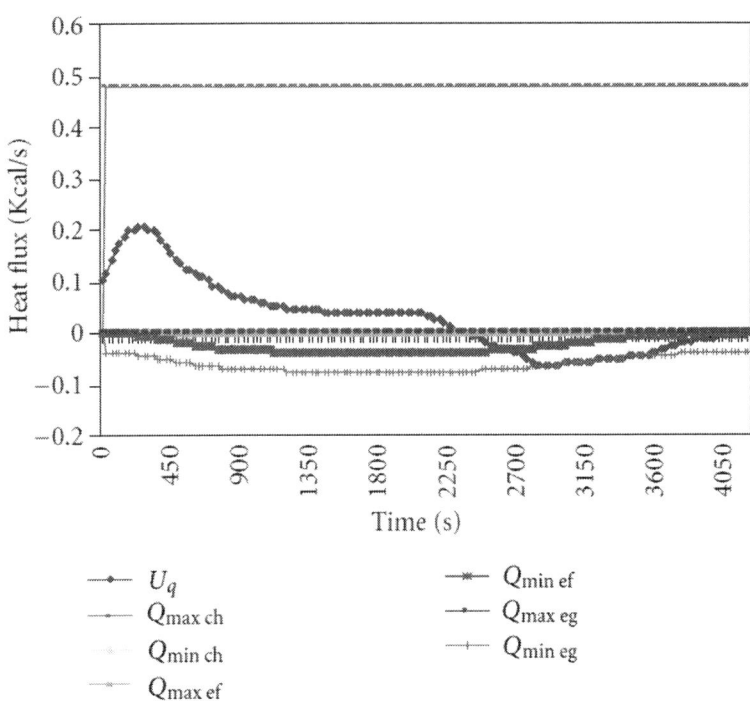

Figure 7: Evolution of thermal flux and limits thermal flux, K = 0.5, Kr = 1.

Figure 5 we can notice that the adjusted set-point temperature (T_{cons}) is correctly followed by the reaction temperature (T_r) and (T_{cdoc}) during almost all the profile of temperature. The only exception, concerns the beginning of the phase of maintenance and the phase of cooling a slight difference appears then of the change of device. This slight difference is also translated by the evolution of the thermal flux (U_q) (Figure 7).

Figure 6 shows the evolution of the manipulated variable (β) which corresponds or in the fraction of the power of heating power calculated by the regulator during use of the electrical resistance or in the fraction of the flow-rate of the mono-fluid calculated during the use of the heat exchanger with cold water either the heat exchanger with glycol water. In the figure, this variable corresponds to the fraction of the heating power during a much of the profile.

Figure 7 shows a regular evolution of the required thermal flux. This variable takes a positive value in time equal has zero to 2400 s ($U_q > Q_{minch}$) allowing the use of the electrical resistance and when it becomes negative ($Q_{minef}<U_q<Q_{maxef}$) in time equal 2430 s to 2610 s it allows the use of PHE which uses the cold water and at times equals 2790 s to 3810 the use of PHE with glycol water, then to 3840 s it takes a positive value. We can notice that U_q plays a role of real supervisor because it permits the change of apparatus only when there are an urgent heating or an urgent cooling without provoking deterioration in the followup of the set-point profile.

Another simulation has been performed for a different set-point profile, with the hydrochloric acid solution was introduced to 1800 s. The simulation consists in a four-step temperature set-point profile.1st phase: heating from 20 to 30°C during 1000 s.2nd phase: constant temperature set-point of 30°C during 5000 s.3rd phase: cooling from 30°C to 20°C during 1000 s.4th phase: constant temperature set-point at 20°C during 500 s.

Results are given in Figures (8–10).

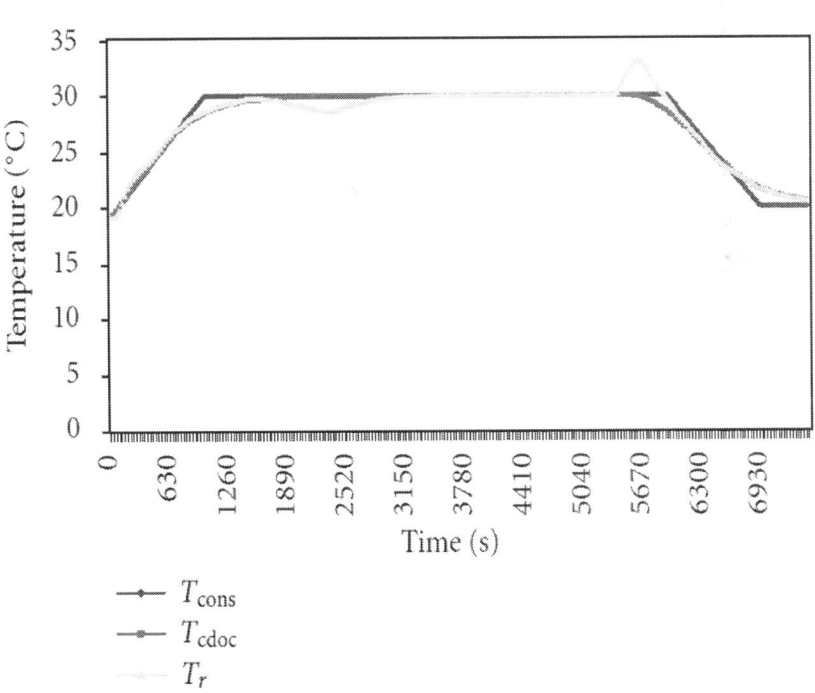

Figure 8: Evolution of temperature (T_{cons}), the adjusted set-point (T_{cdoc}), and reaction mixture (T_r), K = 1, K_r = 0.5.

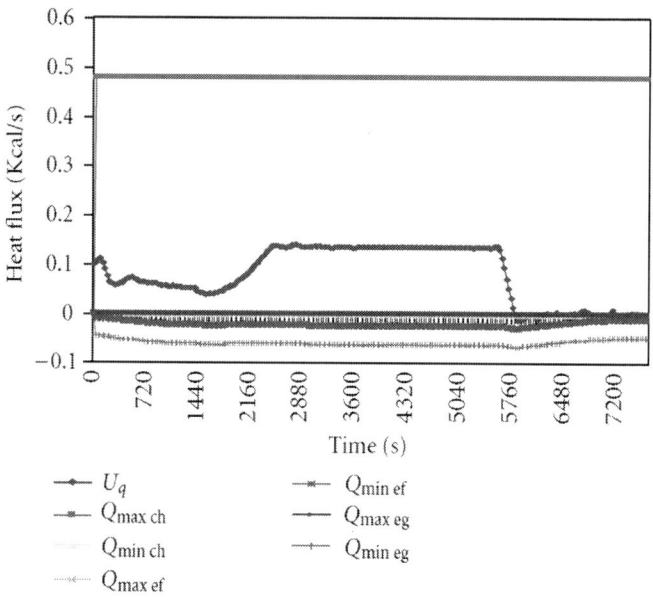

Figure 9: Evolution of thermal flux and limits thermal flux, K = 1, K_r = 0.5.

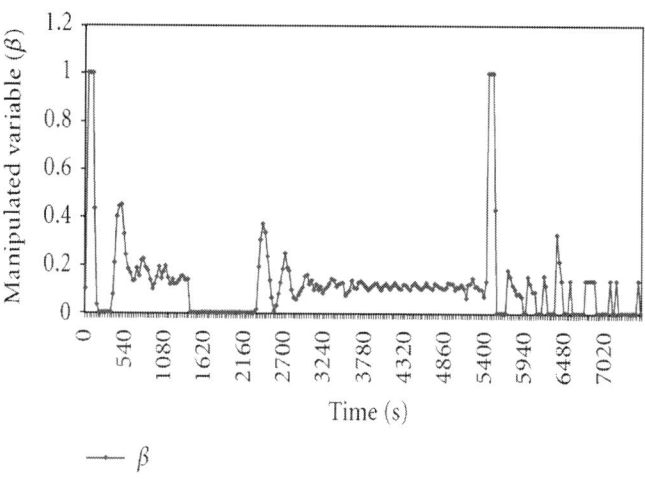

Figure 10: Evolution of the manipulated variable (β), K = 1, K_r = 0.5.

Figure 8 show that at the beginning of the feeding the reaction mixture temperature (T_r) exceeds the set-point temperature (T_{cons}) and (T_{cdoc}). The regulation system reacts always by commuting of the electrical resistance as indicated on Figure 9, and also represents this disturbance translated by the oscillations of the manipulated variable (β) (Figure 10). At the end of feeding, we observe one under overtaking of 1°C approximately, which are due to the fact that the temperature of the mono-fluid is close of 30°C and the system late to return on the landing.

Figure 9 we can observe that the thermal flux in time 5940 s is negative in the range of use of the PHE with cold water and 5970 s until 6000 s the use of the PHE with glycol water then we observe afterward the quality of thermal behavior becomes correct again.

To demonstrate, once again, the performance of the strategy of supervision another simulation has been performed for set-point profile of an exothermic reaction of mononitration with a desired constant temperature of 30°C for various conditions.1st phase: heating from 20 to 30°C during 1000 s.2nd phase: constant temperature set-point between 1000 s and 14800 s with feeding.3rd phase: cooling from 30°C to 20°C during 1000 s.4th phase: constant temperature set-point at 20°C during 500 s

Results are given in Figures (11, 12, and 13).

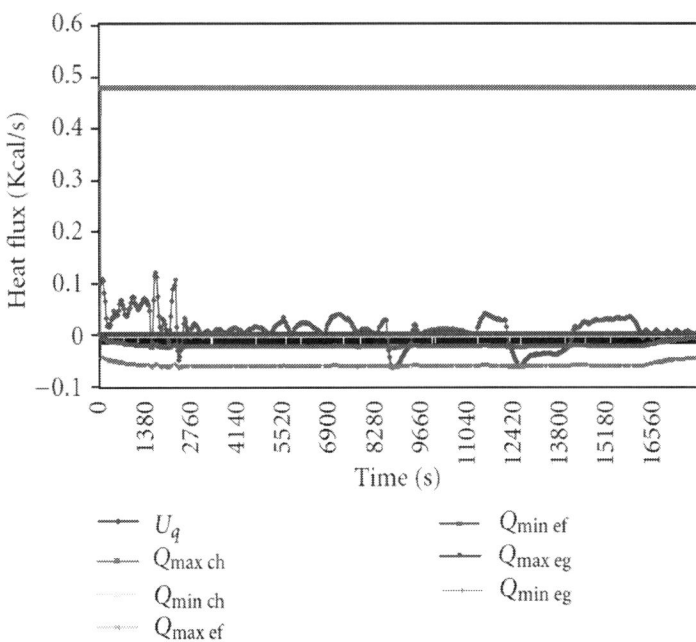

Figure 11: Evolution of thermal flux and limits thermal flux, $K = 0.5$, $K_r = 0.5$.

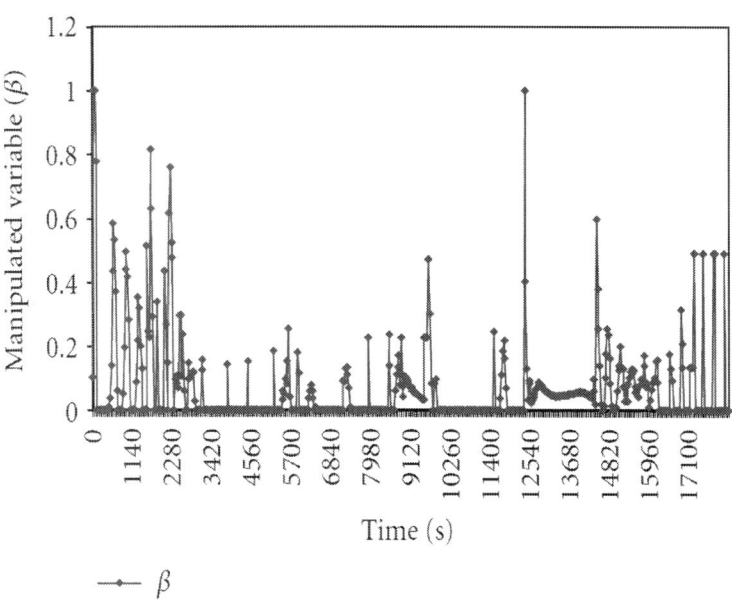

Figure 12: Evolution of the manipulated variable (β), K = 0.5, K_r = 0.5.

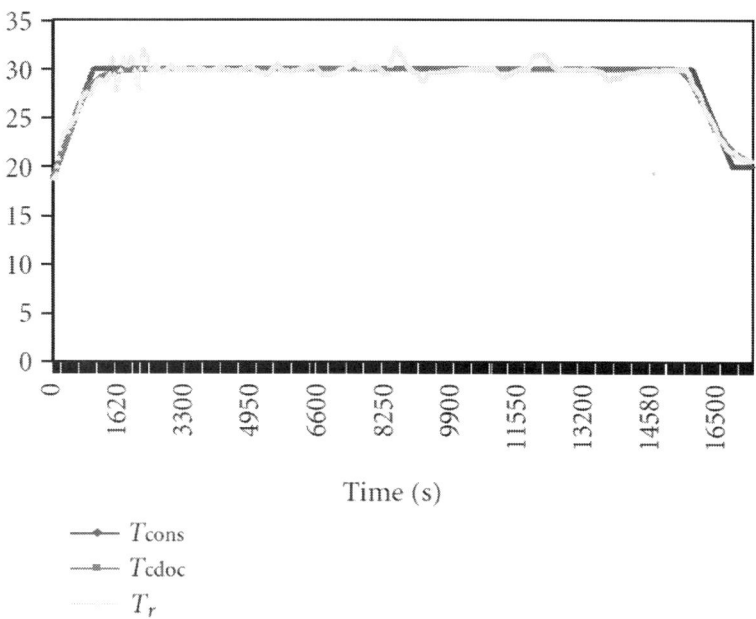

Figure 13: Evolution of temperature (T_{cons}), the adjusted set-point (T_{cdoc}) and reaction mixture (T_r), K = 0.5, K_r = 0.5.

Figures (11–13) show at the time of the introduction of feeding, which is then accompanied by a thermal bad behavior characterized by the oscillations of the thermal flux, the manipulated variable ( ), and the temperature of the reaction mixture, and that for the duration of feeding until the end of the feeding. These oscillations are due to the fact that we did not take into account the release of heat in the model of supervision what generated afterward problems of behavior which are not easy to master.

In addition, if you want the results justified physically by noting that for Figures 5, 6 and 7 obtained for K = 0.5 and $K_r = 1$, the control quality is very good throughout the set-point profile.

Figures 8 to 10 obtained for K = 1 and $K_r = 0.5$, show that one on the other hand has a little unstable behavior for the values K = 0.5 and KR = 0.5 one obtains a very unstable behavior as indicated on Figures 11 to 13.

Analysis of these results shows a value of 1 for the static gain K_r is essential for the master controller is effective while for the regulator responsible for monitoring a value of K in the range from 0.5 to 1 works well.

Note that the controller is the most critical PFC1, it acts as a master controller while PFCM is used for supervision, and its role becomes crucial only in the stages of change of device.

CONCLUSIONS

A new methodology of control and supervision has been developed. It is based on choosing as manipulated variable the thermal flux to be exchanged between the reaction mixture and the jacket. This required flux calculated by the master controller is then used in model based supervisory structure which, according to the limit capacities of the different fluid configurations. In this paper, this methodology has been applied to a mono-fluid heating/cooling system.

The application of such a methodology to pilot-plant batch reactor has been represented. It turned out that at least good basic modeling of the process in necessary. Thus the choice of internal model is crucial importance for (PFC), since the capacity of prediction constitutes the base of all the specifications of performance. However, it is necessary to know the parameters related to the reaction with a certain degree of accuracy. Even if the batch reactor must remain polyvalent, the reactions were generally subject to a certain number of studies. Otherwise, simulation studies have shown when the heat released by the chemical reaction was considered, the control system permits a better tracking of the set-point temperature profile. This strategy is not limited to a mono-fluid heating/cooling system presented in this paper, but may be extended to any system whatever the number of fluids. Therefore the supervision technique can be considered as an appropriate solution for the temperature control of batch reactors, and the developed control strategy can be implemented to control industrials batch reactors.

REFERENCES

1.   M. R. Juba and J. W. Hammer, "Progress and challenges in batch process control," in Proceedings of the Chemical Process Control (CPC '86), vol. 3, pp. 139–183, Elsevier, 1986.

2.   L. D. Schmidt, The Engineering of Chemical Reactions, Oxford University Press, New York, NY, USA, 2005.

3.   M. Friedrich and R. Perne, "Design and control of batch reactors—an industrial viewpoint," Computers and Chemical Engineering, vol. 19, supplement, pp. S357–S368, 1995.

4.   P. Dostál, V. Bobál, and F. Gazdoš, "Simulation of nonlinear adaptive control of a continuous stirred tank reactor," International Journal of Mathematics and Computers in Simulation, vol. 5, no. 4, pp. 370–377, 2011.

5.   K. Preu , M. V. Le Lanne, and G. Anne-Archard, "Supervisory temperature control of batch reactors: application to industrial plant," in Proceedings of the International Symposium on Advanced Control of Chemical Processes (ADCHEM ‹00), vol. 1, pp. 225–230, 2000.

6.   H. Bouhenchir, M. Cabassud, M. V. Le Lann, and G. Casamatta., "A heating/cooling management to improve controllability of batch reactor equipped with a monofluid heating/cooling system," inProceedings of the 10th European Symposium on Computer Aided Process Engineering, pp. 601–606, Sauro Pierucci, Florence, Italy, 2000.

7. F. Xaumier, M. V. Le Lann, M. Cabassud, and G. Casamatta, "Experimental application of nonlinear model predictive control: temperature control of an industrial semi-batch pilot-plant reactor," Journal of Process Control, vol. 12, no. 6, pp. 687–693, 2002. · ·

8. H. Bouhenchir, M. Cabassud, and M. V. Le Lann, "Predictive functional control for the temperature control of a chemical batch reactor," Computers and Chemical Engineering, vol. 30, no. 6-7, pp. 1141–1154, 2006. · ·

9. P. Dostál, J. Vojt šek, and V. Bobál, "Simulation of adaptive control of a continuous stirred tank reactor," in Proceedings of the 23rd European Conference on Modeling and Simulation (ECMS ‹09), pp. 625–630, Madrid, Spain, 2009.

10. P. Dostál, V. Bobál, and F. Gazdoš, "Adaptive control of nonlinear processes: continuous-time versus delta model parameter estimation," in Proceedings of the IFAC Workshop on Adaptation and Learning in Control and Signal Processing (ALCOSP ‹04), pp. 273–278, Yokohama, Japan, 2004.

11. M. Kubal ík and V. Bobál, "Self-tuning control of continuous-time systems," WSEAS Transactions on Systems and Control, vol. 5, no. 10, pp. 802–813, 2010. ·

12. R. Berber, "Control of batch reactors: a review," in Methods of Model Based Process Control, R. Berber, Ed., pp. 459–493, Kluwer Academic Publishers, Dordercht, The Netherlands, 1995.

13. Z. Louleh, M. Cabassud, and M. V. Le Lann, "A new strategy for temperature control of batch reactors: experimental application," Chemical Engineering Journal, vol. 75, no. 1, pp. 11–20, 1999. · ·

14. R. Matusu and R. Prokop, "Experimental verification of design methods for conventional PI/PID controllers," Wseas Transactions on Systems and Control, vol. 5, no. 5, pp. 269–280, 2010. ·

15. G. Marroquin and W. L. Luyben, "Experimental evaluation of nonlinear cascade controllers for batch reactors," Industrial and Engineering Chemistry Fundamentals, vol. 11, no. 4, pp. 552–556, 1972. ·

16. R. W. Chylla and D. R. Haase, "Temperature control of semibatch polymerization reactors," Computers and Chemical Engineering, vol. 17, no. 3, pp. 257–264, 1993. ·

17. H. Bouchenchir, M. Cabassud, M. V. Le Lann, and G. Casamatta, "A general simulation model and a heating/cooling strategy to improve controllability of batch reactors," Trans IChemE, vol. 79, no. 6, pp. 641–654, 2001. ·

18. M.-V. Le Lann, M. Cabassud, and G. Casamatta, "Adaptive model predictive control," in Methods of Model Based Process Control, R. Berber, Ed., pp. 426–458, Kluwer Academic Publishers, Dordrecht, The Netherlands, 1995.

19. J. A. D. Rodrigues, E. C. V. Toledo, and R. Maciel Filho, "A tuned approach of the predictive-adaptive GPC controller applied to a fed-batch bioreactor using complete factorial design," Computers and Chemical Engineering, vol. 26, no. 10, pp. 1493–1500, 2002. · ·

20. B. Ettedgui, Nonlinear predictive control of batch reactors in fine chemistry, Ph.D. thesis, ENSIGC, INP Toulouse, 1999.

21. A. Astolfi, D. Karagiannis, and R. Ortega, Nonlinear and Adaptive Control with Applications, Springer, London, UK, 2008.

22. P. Ioannou and B. Fidan, Adaptive Control Tutorial, SIAM, Philadelphia, Pa, USA, 2006.

23. M. Huba and M. Ondera, "Simulation of nonlinear control systems represented as generalized transfer functions," in Proceedings of the European Control Conference, pp. 1444–1449, Budapest, Hungary, 2009.

24. C. Bányász and L. Keviczky, "A simple PID regulator applicable for a class of factorable nonlinear plants," in Proceedings of the American Control Conference, pp. 2354–2359, Anchorage, Alaska, May 2002.

25. H. Vallery, M. Neumaier, and M. Buss, "Anti-causal identification of Hammerstein models," inProceedings of the European Control Conference, pp. 1071–1076, Budapest, Hungary, 2009.

26. C. Lupu, D. Popescu, and A. Udrea, "Real-time control applications for nonlinear processes based on adaptive control and the static characteristic," Wseas Transactions on Systems and Control, vol. 3, pp. 607–616, 2008.

27. P. Dostál, F. Gazdoš, V. Bobál, and J. Vojt šek, "Adaptive control of a continuous stirred tank reactor by two feedback controllers," in Proceedings of the 9th IFAC Workshop Adaptation and Learning

in Control and Signal Processing (ALCOSP ‹07), pp. 1–6, Saint Petersburg, Russia, 2007.

28. Z. Louleh, M. Cabassud, M. V. Le Lann, A. Chamayou, and G. Casamatta, "A new heating-cooling system to improve controllability of batch reactors," Chemical Engineering Science, vol. 51, no. 11, pp. 3163–3168, 1996. ·

29. Z. Louleh, Modeling and control of batch reactors by thermal flow analysis, Ph.D. thesis, INP Toulouse, France, 1996.

30. H. Bouhenchir, Implementation of predictive control for driving a thermal batch reactor equipped with a single fluid, Ph.D. thesis, INP Toulouse, 2000.

Chapter 5

Thermo-Hydrodynamics of Core-Annular Flow of Water, Heavy Oil and Air Using CFX

Antonio José Ferreira Gadelha[1], Severino Rodrigues de Farias Neto[1], Ramdayal Swarnakar[1], and Antonio Gilson Barbosa de Lima[2]

[1]Department of Chemical Engineering, Center of Science and Technology, Federal University of Campina Grande, Campina Grande, Brasil

[2]Department of Mechanical Engineering, Center of Science and Technology, Federal University of Campina Grande, Campina Grande, Brasil

ABSTRACT

The transport of heavy and ultra-viscous oil employing the core-flow technique has been increasing recently, because it provides a greater reduction of the pressure drop during the flow. In this context, the effect of temperature and the presence of gas on the thermo-hydrodynamics of a three-phase water-heavy oil-air flow in a horizontal pipe under the influence of gravity and drag forces, using the commercial software

ANSYS CFX, have been evaluated. The standard $-$ turbulence model, the mixture model for heavy oil-water system and the particle model for heavy oil-gas and water-gas systems, were adopted. Results of velocity, volume fraction, pressure and temperature fields of the phases present along the pipe are presented and discussed. It has been found that the presence of the air phase and the variation in the temperature affect the behavior of annular flow and pressure drop.

INTRODUCTION

The worldwide heavy oil reserves are estimated to be 3 trillion barrels [1], while light oil reserves have shown a progressive decline in the last decade. This fact leads to an increased economic interest for the reserves of heavy oil which consequently stimulates further research to make production of heavy oil economically feasible. Currently, Brazil is one of the important worldwide producers of oil, and that most of the Brazilian reserves consist of heavy oil in deep waters, generating technical difficulties in exploitation of such resources. Heavy oil is considered to have API between 10 and 20, a density greater than 0.90 g/ml, a viscosity between 10 cP and 100 cP at reservoir conditions and from 100 cP to 10,000 cP at surface conditions [2]. At present, these oils do not have significant economic value due to the low concentration of smaller chain hydrocarbons. However, with the decline of production of light oil, the importance, and consequently the price of these energy sources are likely to increase. The major obstacle of utilizing heavy oil is its relatively high viscosity, which makes it difficult to transport and higher density which increases the cost of refining. Thus, the transportation of heavy and ultraviscous oil is a main technological challenge in the petroleum industry. This fact is related to the high pressure drop or friction due to viscous effects of this type of oil during its flow.

According to Trevisan [3], because of unfavorable characteristics of heavy oils, their transport from the production areas to the processing and refining plants is the biggest obstacle encountered for the production of heavy oils. The author also mentions that the alternatives currently used are to transport by truck or heated pipeline. However, these methods are very expensive and are applicable only for short distances. For efficient transportation at considerable distances, it is necessary to use conventional pipelines, but most of these pipelines have viscosity specifications lower than 0.1 Pa·s, which is not true for heavy oils. In order to overcome the difficulties inherent in the production and transport of heavy and ultra-viscous oil, several techniques have been used, so that a decrease in pressure drop during the flow can be provided, thereby a reduction in the viscosity effect of the fluids present can occur. Among the techniques used one can mention: adding heat to the system, diluting the heavy crude oil with a lighter one and forming emulsions using emulsifying agents. However, each of these alternatives has limitations in their use, both technical and economic.

One technique that has higher efficiency compared to other methods is the core-flow technique. This technique consists in injecting a less viscous liquid, usually water, adjacent to the pipeline wall. This prevents the contact of oil with the inner side of the pipeline. It results in a greater reduction in the pressure drop of the flow and consequently in a reduction in the transport cost of such oil. A drawback of this technique is when the oil comes in contact with the inner wall of the pipeline during transport. This may cause a large increase in the system pressure, which can result in a serious damage to the transportation system and the environment [4].

The most important feature of the core-flow technique is that it does not modify the viscosity of the oil but changes the flow pattern and reduces friction during the transport of very viscous products, such as heavy oil. This reduction in friction also causes a reduction in the longitudinal pressure drop and consequently, a reduction in pumping costs. Several papers have reported research works related to the improvement of the core-flow technique [4-9].

However, in oil production, oil and water rarely flow separately and a gas fraction is generally present, which is characterized as a multiphase flow. Such flow can be defined as a system in which fluid components are immiscible and separated by interfaces. The occurrence of multiphase flow in the oil industry is very common in the units of production, transportation and processing of hydrocarbons of an oil field.

Thus, it is important to analyze the influence of the presence of a third phase (gas) in the annular flow (water-oil) with respect to the pressure drop. In order to verify three-phase flow characteristics some experimental studies have been performed [10-13].

Bannwart et al. [10] studied the pressure drop and flow patterns of a three phase flow observed in a glass tubing with a diameter of 2.84 cm containing heavy oil (3.4 Pa·s and 970 kg/m³ at 20°C), water and air under several combinations of individual compositions and tube inclination (horizontal, vertical and inclined). In their study nine flow patterns were verified. According to these authors, when compared to the two-phase flow of heavy oil-water only, the presence of gas increases considerably the mixture velocity and consequently the pressure drop is increased.

Poesio et al. [12] made an experimental study related to the core-annular flow with the aim of providing a new database for the three-phase (ultra-viscous oil, water and air) flow and to propose a simple model for the determination of the pressure drop. They observed the effect of the air injection on the pressure drop of the annular liquid-liquid flow and noticed an error less than ±15% of measured value in relation to that of the proposed model.

Strazza et al. [13] presented an experimental study of the three-phase flow using water, air and high viscosity oil. The attention was focused on the effect of the gas presence in the core-annular liquid-liquid flow. The experimental flow map, obtained by them, showed that the increase in gas flow breaks up the integrity of the oil core, resulting in a chaotic flow regime. Values for the pressure drop were compared with the proposed theoretical model for the three-phase core flow. The difference between the experimental and the predicted pressure drop was of more or less 20%.

In this background, the objective of this work is to study numerically the three-phase annular flow of heavy oil, water and air (core-flow), at different conditions of temperature and volume fraction of the air.

MATHEMATICAL MODELING

Physical Domain of Study

The study domain consists of a 3 meter long and 2.84 cm inner diameter horizontal tube, in which the flow of heavy oil-water-gas takes place. This is shown in Figure 1.

Computational Domain

To study numerically the annular flow behavior of oilwater in the presence of gas, it is necessary to represent the geometry or domain of study in a computational domain or mesh. For this a mesh of hexahedral structured elements was employed. This mesh was made with the ICEM-CFD commercial package available in ANSYS CFX.

To draw up the computational domain, initially, a tube with two inlets, one annular for water injection and another circular for the oil together with gas, was created. This can be seen in Figure 1. The ring or the annular space between tube wall and the oil core has a thickness of 1.7 mm. The mesh used in the present work consisted of 464,000 hexahedral elements.

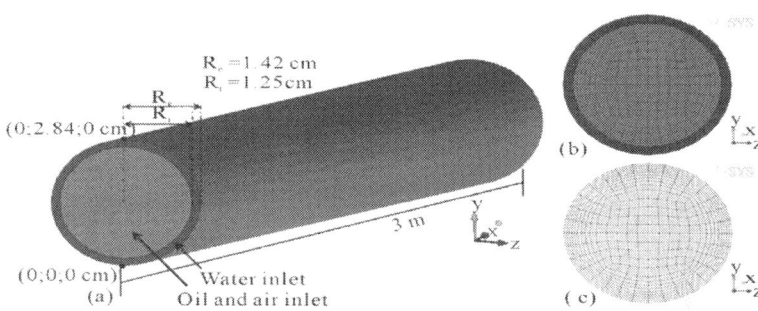

Figure 1: (a) Tube dimensions and details of water and oil inlet sections. (b) Details of the expanded mesh at the inlet region and (c) at the outlet region.

Mathematical Model

To study the three-phase flow within a horizontal pipe following conditions have been considered:

- Incompressible and steady state flow;
- No chemical reactions;
- Existence of gravitational and drag effects;
- The viscosities of water, gas and ultra-viscous heavy oil are functions of temperature;

- There is no interfacial mass transfer between water, oil and gas phases;

Thus, the conservation equations of mass, momentum and energy applied to the multiphase flow are reduced to:

Conservation Equations

Mass Conservation Equation

$$\nabla \bullet \left(r_\alpha \rho_\alpha \vec{U}_\alpha \right) = 0$$

(1)

Where r_α, ρ_α^e $\bar{U}_\alpha$ correspond to the volume fraction, density and velocity vector of phase α, respectively.

Momentum Conservation Equation

$$\nabla \bullet \left[r_\alpha \left(\rho_\alpha \vec{U}_\alpha \otimes \vec{U}_\alpha \right) \right] = -r_\alpha \nabla p_\alpha +$$
$$\nabla \bullet \left\{ r_\alpha \mu_\alpha \left[\nabla \vec{U}_\alpha + \left(\nabla \vec{U}_\alpha \right)^\Gamma \right] \right\} + \vec{S}_{M_\alpha} + \vec{M}_\alpha$$

(2)

where p is pressure, $\vec{S}_{M_\alpha}$ represents a term for the external forces acting on the system per unit volume and $\vec{M}_\alpha$ describes the total forces per unit volume (interfacial drag force).

In the mixture model, available in ANSYS CFX, the total forces per unit volume only consider the interfacial forces (drag) and are given by:

$$\vec{M}_\alpha = \vec{M}_{\alpha\beta} = C_D A_{\alpha\beta} \rho_{\alpha\beta} \left| \vec{U}_\beta - \vec{U}_\alpha \right| \left(\vec{U}_\beta - \vec{U}_\alpha \right)$$

(3)

where C_D is the drag coefficient and the sub-indexes α and β correspond to the phases α and β present in the flow.

In Equation (3), $A_{\alpha\beta}$ corresponds to the interfacial contact area between the phases α and β which is given by:

$$A_{\alpha\beta} = \frac{6r_\beta}{d_\beta}$$

(4)

where in r_β and d_β represent the volume fraction and the diameter of the particle of the phase β, respectively. Adopting the particle model, α represents the continuous phase (heavy oil or water) and β being the dispersed phase (gas).

For the mixture model the interfacial contact area between phases α and β is given by:

$$A_{\alpha\beta} = \frac{r_\alpha \cdot r_\beta}{d_{\alpha\beta}}$$

(5)

where $d_{\alpha\beta}$ is the length of the mixture.

Energy Conservation Equation

$$\nabla \bullet \left[r_\alpha \left(\rho_\alpha \vec{U}_\alpha h_\alpha - \lambda_\alpha \nabla T_\alpha \right) \right] = Q_\alpha$$

(6)

where h_α, λ_α and T_α describe static enthalpy, thermal conductivity and temperature respectively of α phase and Q_α describes heat transfer to the α phase through the interfaces with the other phases, which is given by:

$$Q_\alpha = \sum_{\beta \neq \alpha} Q_{\alpha\beta}$$

(7)

where:

$$Q_{\alpha\beta} = -Q_{\beta\alpha} \Rightarrow \sum_{\alpha} Q_\alpha = 0$$

(8)

The heat transfer through the boundary is usually described in terms of a coefficient of overall heat transfer, $h_{\alpha\beta}$, which is the amount of heat energy through a unit area per unit time per unit of temperature difference between the phases.

Thus, the rate of heat transfer, $Q_{\alpha\beta}$, per unit of time through the phase interfacial boundary area, per unit volume $A_{\alpha\beta}$, of phase β to phase α is given by:

$$Q_{\alpha\beta} = h_{\alpha\beta} A_{\alpha\beta} \left(T_\beta - T_\alpha \right)$$

(9)

Many times it is convenient to express the coefficient of heat transfer in terms of the dimensionless Nusselt number, defined by Equation (10):

$$h = \frac{\lambda Nu}{d}$$

(10)

In the particle model the thermal conductivity (l) is considered as being the thermal conductivity of the continuous phase, and the length d is considered to be the diameter of the dispersed phase. So it can be written as:

$$h_{\alpha\beta} = \frac{\lambda_\alpha Nu_{\alpha\beta}}{d_\beta}$$

(11)

Turbulence Model

As turbulence model, the standard $K - \varepsilon$ model was used, where it is assumed that the Reynold's tensors are proportional to the mean velocity gradients, with the proportionality constant being characterized by turbulent viscosity (idealization known as Boussinesq hypothesis).

The characteristic of this type of model is that two transport equations modeled separately are solved for the turbulent length and the time scale or solved for any two linearly independent combinations of them. The transport equations for the turbulent kinetic energy, K, and turbulent dissipation rate, ε, respectively are:

$$\nabla \bullet \left\{ r_\alpha \left[\rho_\alpha \vec{U}_\alpha \kappa_\alpha - \left(\mu + \frac{\mu_{t\alpha}}{\sigma_k} \right) \nabla \kappa_\alpha \right] \right\} = r_\alpha \left(G_\alpha - \rho_\alpha \varepsilon_\alpha \right)$$

(12)

$$\nabla \bullet \left\{ r_\alpha \rho_\alpha \vec{U}_\alpha \varepsilon_\alpha - \left(\mu + \frac{\mu_{t\alpha}}{\sigma_\varepsilon} \right) \nabla \varepsilon_\alpha \right\} = r_\alpha \frac{\varepsilon_\alpha}{\kappa_\alpha} \left(C_1 G_\alpha - C_2 \rho_\alpha \varepsilon_\alpha \right)$$

(13)

where G_α is turbulent kinetic energy generated in the phase α, C_1 and C_2 are empirical constants. Also, in this equation, ε_α the dissipation rate of turbulent kinetic energy of phase α and K_α the turbulent kinetic energy for phase α , are defined by:

$$\varepsilon_\alpha = \frac{c_\mu q_\alpha^3}{l_\alpha}$$

(14)

$$K_\alpha = \frac{q_\alpha^2}{2}$$

(15)

where l_α is the length of spatial scale, q_α is the velocity scale, $c_\mu = 0.09$ is an empirical constant.

The variable $\mu_{t\alpha}$ is the turbulent viscosity and is given by the equation:

$$\mu_{t\alpha} = c_\mu \rho_\alpha \frac{K_\alpha^2}{\varepsilon_\alpha}$$

(16)

where, the constants used in the above equations are: $C_1=1.14; C_2=1.92;\ \sigma_k = 1.0\ \text{and}\ \sigma_\varepsilon = 1.3$.

Boundary Conditions

- In the annular section referring to water inlet a prescribed and not null value, for the axial velocity component and volume fraction of water in the x direction, was adopted such that:

$$R_i < y < R_e, \text{ at } x = 0; \begin{cases} u_w \neq 0 \\ v_w = w_w = 0 \\ u_o = v_o = w_o = 0 \\ u_g = v_g = w_g = 0 \\ r_o = r_g = 0 \\ r_w = 1 \\ T = T_w \end{cases}$$

Where u, v_e v correspond to the velocity vector components in the x, y and z directions, in this order, the sub-indices w, o and g represent water, oil and gas phases respectively, and T the temperature.

- In the core section referring to oil inlet a prescribed and not null value, for the axial velocity component and volume fraction of oil and air in the x direction, was adopted such that:

$$0 < y < R_i, \text{ at } x = 0; \begin{cases} u_o = u_g \neq 0 \\ v_o = w_o = 0 \\ v_g = w_g = 0 \\ u_w = v_w = w_w = 0 \\ r_o = 0.95 \\ r_g = 0.05 \\ r_w = 0 \\ T = T_o = T_g \end{cases}$$

- At the borders referring to the tube wall the condition of non-slip was considered, namely:

$$y = R_e, \text{ at } 0 \le x \le L; \begin{cases} u_w = v_w = w_w = 0 \\ u_o = v_o = w_o = 0 \\ u_g = v_g = w_g = 0 \\ T = T_p = 288K \end{cases}$$

- At the outlet section (x = L), a constant mean pressure p_{est} = 101,325 Pa, was prescribed, where L is the length of the tube.

In the present work a root mean square (RMS) residue equal to 10^{-7} kg/s was considered as the convergence criterion. The thermo-physical properties of the fluids used in the simulation are presented in Tables 1 and 2. Table 3 summarizes all the cases of this study.

RESULTS AND DISCUSSION

Influence of the Air Phase on Flow Hydrodynamics

Figures 2 and 3 depict the superficial velocity profiles for oil and water, respectively. The two-phase flow (oilwater) refers to the case 01, and the three-phase (oil-water-air), refers to the case 04 (Table 3), at four positions of the axial direction, Z = 0 m and were obtained under the same conditions.

By observing the velocity profiles in Figures 2 and 3, it can be noted that the introduction of the air phase in two-phase flow (water-oil), causes a significant change in fluid-dynamics of water and oil phases. It affects the velocity gradient in the lower region of the pipe due to the increase in water flow in this region. A similar behavior was observed independently by [10-13]. This effect can be better seen in Figures 4(a) and (b), where the behavior of the volumetric fraction field of oil in the pipe, at different axial positions, is illustrated. Here it is clearly perceived that the presence of air creates a greater elevation of the oil core and, thus increases the area occupied by annular section of water in the lower region of the pipe, close to the wall. Therefore, it can be said that the divergence between buoyancy and lubrication forces are more intense in three phase flow than in two-phase flow.

Table 1: Thermo-physical properties of the fluids (25°C) used in the simulations

Properties	Water	Heavy Oil	Air	Source
Density (kg/m3)	997,200	971,000	0.778	[14]
Specific heat (J/kg·K)	4181,700	1800,000	1025,766	[15]
	Water/Oil	Water/Air	Oil/Air	
Surface tension (N/m)	0.067	0.0725	0.026	[14]

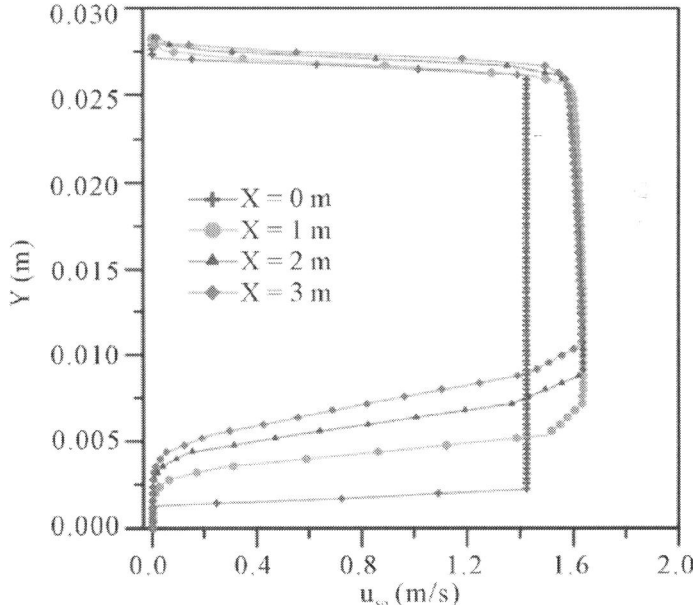

(a)

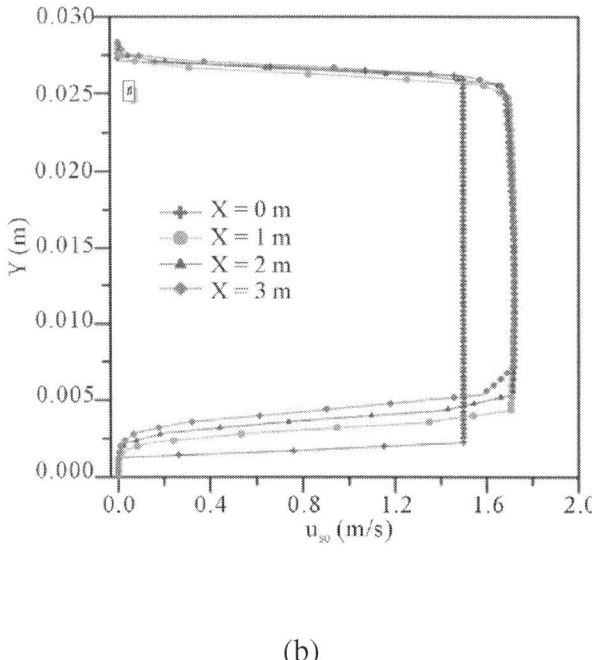

(b)

Figure 2: Superficial velocity profiles of the oil: (a) three-phase (water-oil-air) flow and (b) two-phase (water-oil) flow.

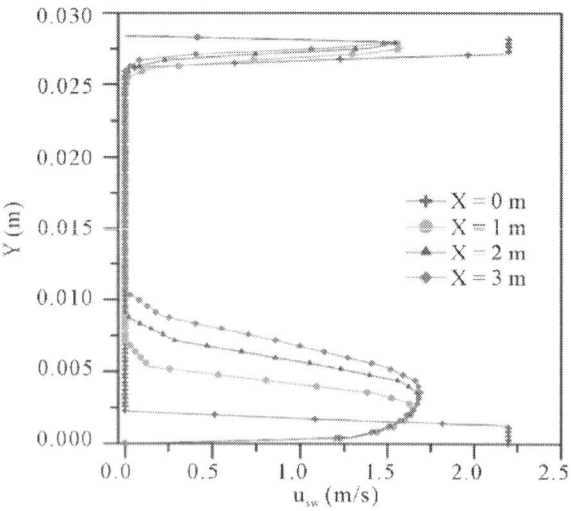

(a)

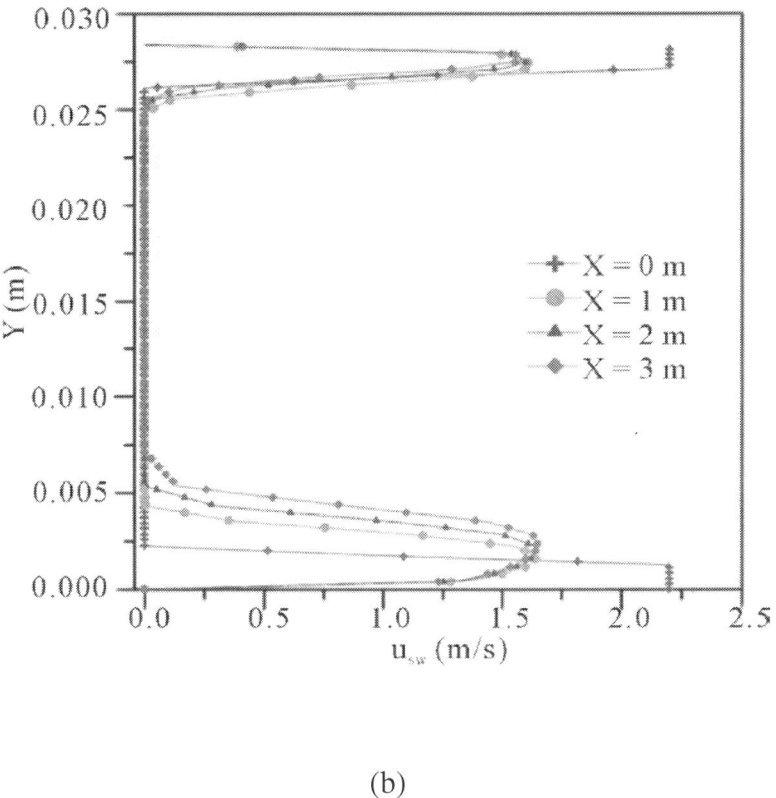

(b)

Figure 3: Superficial velocity profiles of the water: (a) Three-phase (water-oil-air) flow and (b) Two-phase (water-oil) flow.

Table 2: Dynamic viscosity of the fluid as a function of temperature (in °C)

Fluid	Dynamic Viscosity (Pa.s)**	Source
Water	$\mu_w = \left(\dfrac{997.2}{2.443299 \times 10^{-2} \times T - 6.153676} \right)$	[16]
Oil	$\mu_o = 0.6402 + 18.9612 \times e^{(-0.074 \times T)}$	[3]
Air	$\mu_g = 2.8 \times 10^{-7} \times T^{0.735476}$	[17]*

*Equation fitted to the data provided by [17]; **Applied in the range of 0°C ≤ T ≤ 100°C.

Effect of Temperature on Flow Hydrodynamics

Figure 5 depicts the superficial velocity profiles, at different temperatures (288.15 K, 303.15 K and 323.15 K) for the cases: 02, 03 and 05, on the axial position equal to 1.0 m and direction Z = 0 m. It is noted that an increase in the temperature of the phases, at the entrance of the pipe, causes a small variation in the oil phase superficial velocity due to the change in the viscosity.

Figure 6 shows the volumetric fraction field of the oil phase at a distance X = 1 m, for different temperatures of the phases at the pipe entrance. It is noticed that as the temperature increases, the oil core level undergoes a small increase. This effect of temperature can be explained in terms of the reduction in the oil viscosity, whereby the resistance to the flow of this fluid, caused by viscous forces, is reduced.

Table 3: Conditions used in the simulations

Case	uw (m/s)	uo (m/s)	ug (m/s)	rw	ro in mixture	rg in mixture	Tw, To and Tg (K)
01 (Two-phase)	2.20	1.50	-	1.00	1.00	-	313.15
02 (Three-phase)	2.20	1.50	1.50	1.00	0.95	0.05	288.15
03 (Three-phase)	2.20	1.50	1.50	1.00	0.95	0.05	303.15
04 (Three-phase)	2.20	1.50	1.50	1.00	0.95	0.05	313.15
05 (Three-phase)	2.20	1.50	1.50	1.00	0.95	0.05	323.15

$^*T_w$ and T_o only.

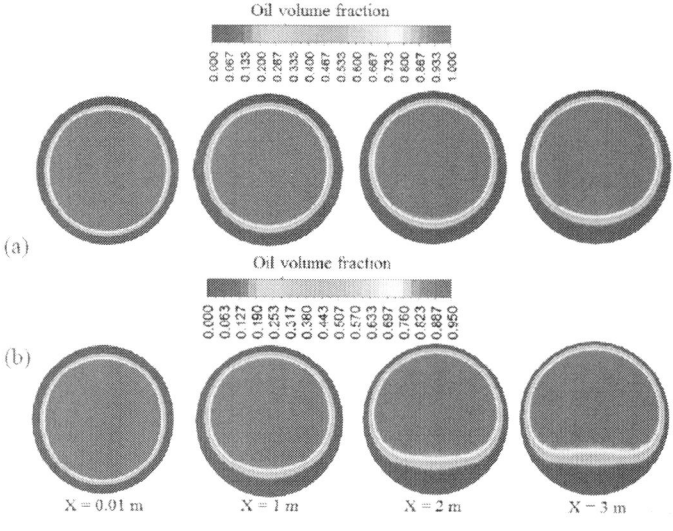

Figure 4: Oil volume fraction field in two-phase flow (a) and three-phase flow (b), at different YZ cross sections along the pipe.

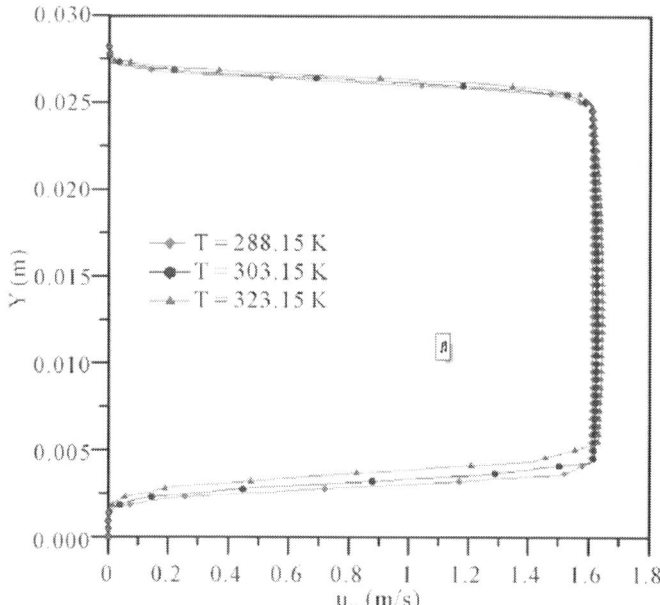

Figure 5: Superficial velocity profiles of the oil at different temperatures (X = 1 m and Z = 0 m).

Effect of Temperature and the Presence of the Gas on Pressure Drop

Table 4 presents the values for the pressure drop as a function of temperature for Cases 01 to 05 (two and three phases). It can be seen that for the three-phase flow (oilwater-gas) an increase in temperature results in a reduction of pressure drop, ΔP, which is more pronounced in the temperature range 288 to 313 K. This fact can be explained due to the decrease in viscosity of oil and water with increase in temperature, which reduces the resistance to the flow in the pipe and thereby causing a decrease in pressure drop. The viscosity of the air increases with increasing temperature, but as it is present in a lesser volumetric fraction, this effect is small compared to other phases (oil-water). Comparing the two-phase flow (Case 01) with the three-phase flow (case 04), it can be noted that the presence of air causes an increase in pressure drop of the flow. Trevisan [3] and Bannwart et al. [10] have verified a similar behavior for the simultaneous flow of heavy oil, water and gas. According to the latter, this is due to the fact that the gas increases the velocity of the fluid. The increase in the velocity also increases the friction factor and hence the pressure drop of the three phase flow.

Profiles and Temperature Fields of the Phases

Figures 7 and 8 show the temperature profiles for water and oil respectively (Case 04) at four axial positions (0, 1, 2 and 3 m) along the pipe It can be seen that the water temperature at the pipe entrance (0 m) is uniform, due to the boundary conditions assumed. One can perceive a temperature decrease, as the fluids move away from the entrance. A strong temperature gradient near the pipe wall, due to the boundary condition adopted, is also observed.

By observing Figure 7, which illustrates the water temperature profile along the tube, it is verified that the water has a greater temperature reduction at the upper region of the pipe. This behavior is due to the fact that, as the water tends to accumulate in the lower region, the film of water formed on top of the flow is relatively thinner, and this film thus undergoes greater influence of the low temperature adopted for the pipe wall.

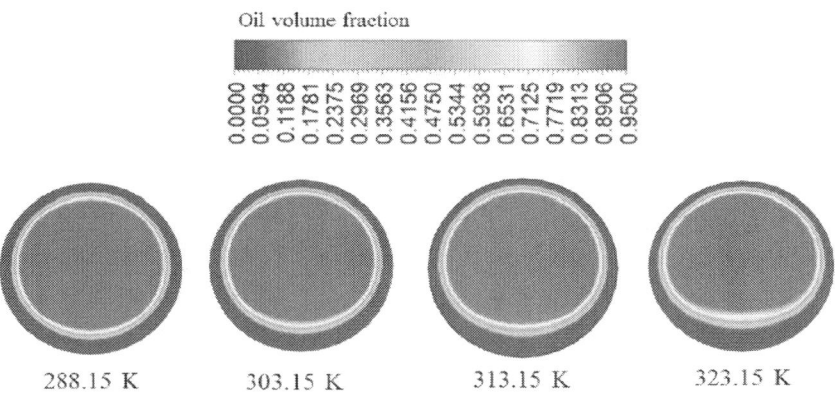

Figure 6: Volumetric fraction field of oil for different temperatures, on YZ plane (transversal section), at 1 m distance from the entrance.

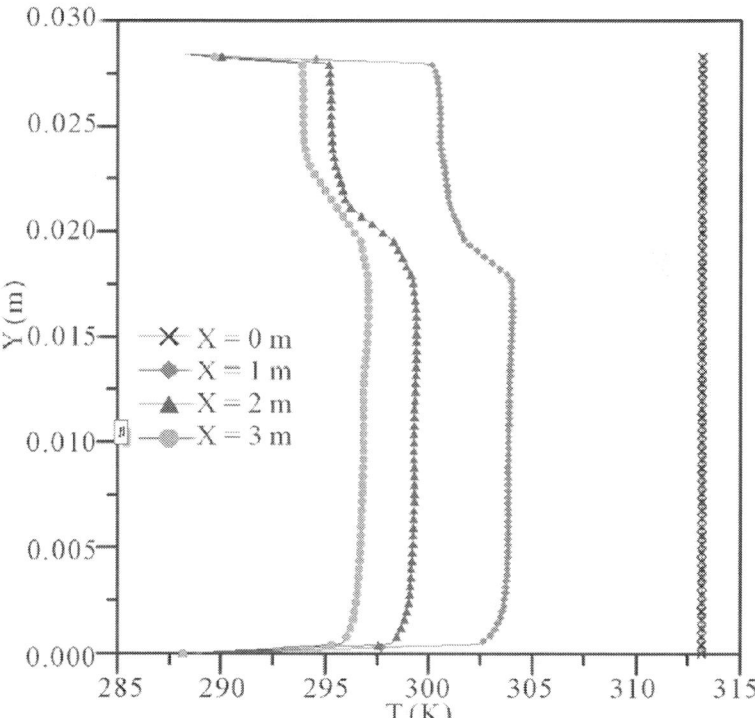

Figure 7: Water temperature profiles at four axial positions (X) along the pipe, at Z = 0 m (Case 04).

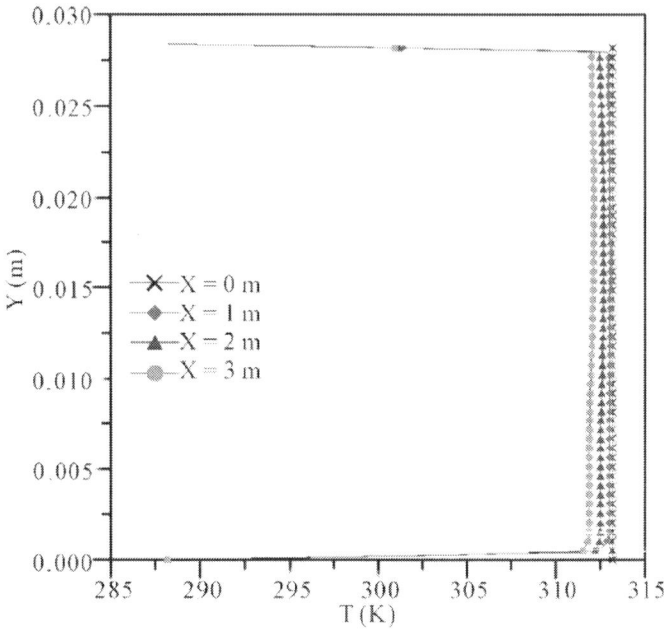

Figure 8: Oil temperature profiles at four axial positions (X) along the pipe, at Z = 0 m (Case 04).

With respect to the oil temperature profile (Figure 8), there is a uniform behavior in the central region of the pipe. However, as one moves away from the inlet section, a small temperature decrease can still be noted. This can be attributed to the heat transfer, since the pipe wall is at a lower temperature than the oil. By comparing Figures 7 and 8, it can be seen that the oil at the pipe exit has a temperature higher than the temperature of the water, which is due to the fact that the water flows next to the wall preventing contact between the oil and the wall, thus water behaves as a thermal insulator.

Figures 9 and 10 present the details of temperature fields of water and oil, at the inlet (Figure 9(a)) and at the outlet (Figure 9(b)) sections near the pipe wall, respectively. It is in this region where the main temperature changes occur. It is possible to visualize the formation of the thermal boundary layer, near the pipe entrance, (Figures 9(c) and (d)), which is due to the temperature difference between the wall and the adjacent fluid. Figures 9 and 10, therefore, corroborate that the temperature distributions in oil and water phases differ in the pipe and that water undergoes a higher decrease in its temperature.

Table 4: Pressure drop per unit length as a function of the temperature of the mixture at the entrance of the pipe

Case	Tw, To, Tg (K)	ΔP (Pa/m)
01 (Two-Phase)	313.15*	1267.50
02 (Three-Phase)	288.15	1561.50
03 (Three-Phase)	303.15	1483.83
04 (Three-Phase)	313.15	1465.88
05 (Three-Phase)	323.15	1461.25

*T_w and T_o only.

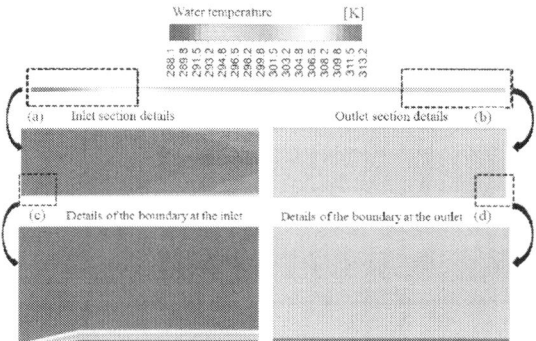

Figure 9: Water temperature field in the XY plane along the pipe (Case 04).

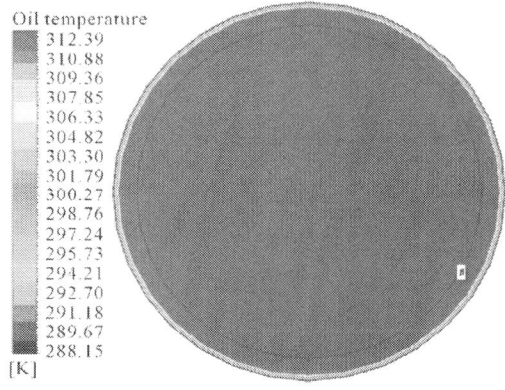

Figura 10: Oil temperature field in the YZ plane at X = 1 m, along the pipe (Case 04).

With respect to the temperature of the air phase, as it is dispersed in the oil core, its temperature profile is similar to the temperature profile of the continuous phase (heavy oil), presenting a uniform behavior in the vicinity of the center of the pipe.

CONCLUSIONS

Based on the results obtained it can be concluded that:

- The utilized mathematical model was able to predict the behavior of a non-isothermal three-phase oil-waterair flow in a horizontal pipe;

- It is found that the position of the oil core in the pipe and the velocity profiles of the phases are affected by the presence of air in the heavy oil-water two-phase flow. However, the core-annular flow pattern maintained its integrity;

- The core-annular flow pattern is maintained even with temperature variation, and that the core of oil and air mixture tends to stratification (to be eccentric) while keeping away from the wall by a formed thin film of water;

- Increasing the temperature of the fluids injected causes a reduction in the flow pressure drop, due to the decrease in the viscosity of oil and water, while the presence of air causes an increase in the pressure drop;

- Temperature profiles of the phases along the pipe are affected by the assumed lower temperature of the pipe wall. It causes a reduction in the temperature of the fluids at the exit of the pipe, and this reduction is greater for annular water which is in contact with the pipe wall.

ACKNOWLEDGEMENTS

The authors of the present work would like to express their thanks to CNPQ, CAPES, FINEP, ANP/UFCG/ PRH-25, PETROBRAS and JBR Engenharia Ltda. (Brazil), for the technical and financial support.

REFERENCES

1. K. C. O. Crivelaro, Y. T. Damacena, T. H. F. Andrade, A. G. B. de Lima and S. R. de Farias Neto, "Numerical Simulation of Heavy oil Flows in Pipes Using the CoreAnnular Flow Technique," WIT Transactions on Engineering Sciences, Computational Methods in Multiphase Flow V, Vol. 63, No. 5, 2009, pp. 193-203.http://dx.doi.org/10.2495/MPF090171

2. R. M. O. Vara, A. C. Bannwart and C. H. M. Carvalho, "Production and Transportation of Heavy Oil by Water Injection," 1st Brazilian Congress R & D in Petroleum and Gas—1st PDPETRO, UFRN-SQB Regional, Natal, 25-28 November 2001 (In Portuguese).

3. F. E. Trevisan, "Flow Patterns and Pressure Drop in Three Phase Horizontal Flow of Heavy Oil, Water and Air," Master's Thesis, Petroleum Science and Engineering, Faculty of Mechanical Engineering, State University of Campinas (UNICAMP), Campinas, 2003 (In Portuguese).

4. A. C. Bannwart, "Modeling Aspects of Oil-Water CoreAnnular Flows," Journal of Petroleum Science and Engineering, Vol. 32, No. 2-4, 2001, pp. 127-143.http://dx.doi.org/10.1016/S0920-4105(01)00155-3

5. G. Ooms and P. Poesio, "Stationary Core-Annular Flow through a Horizontal Pipe," Physical Review E, Vol. 68, No. 066301, 2003, pp. 1-7.http://dx.doi.org/10.1103/PhysRevE.68.066301

6. A. Bensakhria, Y. Peysson and G. Antonini, "Experimental Study of the Pipeline Lubrication for Heavy Oil Transport," Oil & Gas Science and Technology, Vol. 59, No 5, 2004, pp. 523-533. http://dx.doi.org/10.2516/ogst:2004037

7. S. Ghosh, T. K. Mandal, G. Das and P. K. Das, "Review of Oil Water Core Annular Flow," Renewable and Sustainable Energy Reviews, Vol. 13, No. 8, 2009, pp. 1957- 1965.http://dx.doi.org/10.1016/j.rser.2008.09.034

8. J. S. de S. Santos, "Numerical Study of Submerged Risers Lubrication to Transport of Heavy Oil," Master's Thesis, Chemical Engineering, Center of Science and Technology, Federal University of Campina Grande (UFCG), Campina Grande, 2009 (In Portuguese).

9.   T. H. F. Andrade, K. C. O. Crivelaro, S. R. de Farias Neto and A. G. B. de Lima, "Numerical Study of Heavy Oil Flow on Horizontal Pipe Lubricated by Water," In: A. Öchsner, L. F. M. da Silva and H. Altenbach, Eds., Materials with Complex Behaviour II: Advanced Structured Materials, Springer-Verlag, Heidelberg, Berlin Heidelberg, Vol. 16, 2012, pp. 99-118. http://dx.doi.org/10.1007/978-3-642-22700-4_6

10.   A. C. Bannwart, O. M. H. Rodriguez, F. E. Trevisan, F. F. Vieira and C. H. M. de Carvalho, "Experimental Investigation on Liquid-Liquid-Gas Flow: Flow Patterns and Pressure-Gradient," Journal of Petroleum Science and Engineering. Vol. 65, No. 1-2, 2009, pp. 1-13. http://dx.doi.org/10.1016/j.petrol.2008.12.014

11.   P. Poesio, G. Sotgia and D. Strazza, "Experimental Investigation of Three-Phase Oil-Water-Air Flow through a Pipeline," Multiphase Science and Technology, Vol. 21, No. 1-2, 2009, pp. 107-122. http://dx.doi.org/10.1615/MultScienTechn.v21.i1-2.90

12.   P. Poesio, D. Strazza and G. Sotgia, "Very-Viscous-Oil/ Water/Air Flow through Horizontal Pipes: Pressure Drop Measurement and Prediction," Chemical Engineering Science, Vol. 64, No. 6, 2009, pp 1136-1142. http://dx.doi.org/10.1016/j.ces.2008.10.061

13.   D. Strazza, D. Chiecchi and P. Poesio, "High Viscosity Oil-Water-Air Three Phase Flows: Flow Maps, Pressure Drops and Bubble Dynamics," 7th International Conference on Multiphase Flow—ICMF, Tampa, 30 May-4 June 2010, pp. 1-7.

14.   F. N. Silva, T. H. F. Andrade, S. R. de Farias Neto and A. G. B. de Lima, "Numerical Study of Three Phase Flow (Water-Heavy Oil-Gas) Type Core Flow in a Connection 'T'," 6th Brazilian Congress of Research and Development at Petroleum and Gas—6th PDPETRO, Florianópolis, 9-13 October 2011 (In Portuguese).

15.   F. P. Incropera, A. S. Lavine and D. P. DeWitt, "Fundamentals of Heat and Mass Transfer," 6th Edition, LTC, Rio de Janeiro, 2008 (In Portuguese).

16.   C. W. S. Santana, E. G. Tôrres and I. de S. Lacerda, "Adjustment Equations for Kinematic Viscosity of Petroleum Products Depending on the Temperature," Proceedings of the 3rd Brazilian Congress of R & D in Petroleum and Gas—3rd PDPETRO, Rio de Janeiro, 2-5 October 2004 (In Portuguese).

17.   F. Kreith and M. S. Bohn, "Principles of Heat Transfer," Editora Edgard Blücher, São Paulo, 1977 (In Portuguese).

Modeling and Analysis of SO$_2$ Emissions under Fast Fluidized Bed Conditions Using One Dimensional Model

Khurram Shahzad[1], Mahmood Saleem[2],
Moinuddin Ghauri[3], Waqar Ali Khan[4],
and Niaz Ahmed Akhtar[2]

[1]Centre for Coal Technology, University of the Punjab, Lahore, Pakistan

[2]Institute of Chemical Engineering & Technology, University of the Punjab, Lahore, Pakistan

[3]Department of Chemical Engineering, Comsats Institute of IT, Lahore, Pakistan

[4]NFC-IEFR, Faisalabad, Pakistan

ABSTRACT

Fluidized bed combustion behavior of coal and biomass is of practical interest due to its significant involvement in heating systems and power plant operations. This combustion behavior has been studied by many experimental techniques along with different kinetic models. In this study, SO_2 emissions have been studied out in a pilot scale test facility of Circulating Fluidized Bed combustor (70 KW) under fast fluidized bed conditions burning coal with Pakistani wheat straw. One dimensional Mathematical model is being developed to predict the SO_2 emissions under different operating conditions like bed temperature, Ca/S molar ratio, solids circulation rate, excess air ratio and secondary to primary air ratio. These parameters are varied to validate the model and encouraging correlation is found between the experimental values and model predictions.

INTRODUCTION

Biomass as an alternative energy source is getting a lot of attention due to the environmental and cost benefits. Globally, attention has been diverted for the replacement of fossil fuels with biomass. In UK, utilization of the fossil fuels will be replaced with renewables by 10% up to 2010 and 20% up to 2020 [1] . CO_2 and SO_2 emissions from coal fired power plants can effectively be reduced by co-firing the CO_2 neutral fuels with coal. About 534.23 million tons of wheat straw is produced worldwide in 2011 [2] . A major portion of the wheat stubble is burned in the field which causes significant environmental and health problems [3] . Reduced SO_2 emissions have been reported during the combustion of coal and Pakistani wheat straw under fast fluidized bed conditions [4] . About 1200 circulating fluidized bed combustion (CFBC) plants with installed capacity of 65 GW_{th} are in operation worldwide [5] . Along with reduction of CO_2, biomass also reduces NO_x, SO_2 and CO emissions in cocombustion with coal [6] -[8] . Gaseous emissions from co-firing have also being reported as the function of operating conditions [4] [9] .

The objective of the present study is to model the CFB rig for the estimation of SO_2 emissions under different operating conditions and compare the model values to the experimental values. The values of bed temperature, Ca/S molar ratio, solids circulation rate, excess air ratio and secondary to primary air were varied to validate the model.

Hannes classified the different types of model based on complexity as global models, one dimensional model, multi-dimensional model (computational fluid dynamics) and scaling and expert systems [10] .

Hartleben introduced the first model for atmospheric and pressurized circulating fluidized beds in which an empirical approach was used for the fluid dynamics and particle size distribution [11] . One dimensional model based on different blocks was also developed for a boiler at Tsinghua University, China [12] . Another dynamic model was developed for the multi solid fluidization in CFB to predict the temperature along the height [13] [14] . Basu introduced a new generalized model with sensitivity analysis based on two zones vertically to find the temperature, sulphur capture and NO$_x$ formation [15] -[17] . Different models had been developed to estimate the concentrations of CO, CO$_2$, and NO$_x$ [18] and to calculate the oxygen concentration, carbon fraction and char size distribution [19] . Another detail modelling was done to calculate the char combustion, temperature distribution, pollutant formation and heat transfer [20] . Haider developed a detailed CFBC model to cover the cyclone and external heat exchanger performance along with fluid dynamics and chemistry of the reactions involved [21] [22] . In Table 1, an overview of some models from the literature is being summarized.

Hannes developed a very detail and comprehensive model for the coal combustion in CFB boiler [10][23] [24] . In this model, sub models were developed for each calculation and then recalled into main program. Sub models covered the fluidization pattern of solid flow, development of the particle size distribution, gas flow, coal conversion reactions, homogeneous and heterogeneous gas reactions and heat transfer mechanisms.

MATERIAL AND METHODS

Experimental Setup

CFB combustor used in this investigation is shown in Figure 1. The system comprised of a riser of 0.152 m i.d. and 6.2 m height, two high efficiency cyclones in series, an external heat exchanger (EHE) and an L-valve. The coal and wheat straw were supplied from gravimetric hoppers with screw feeders coupled with variable speed motors.

Table 1: Comparison of some models given in literature

	Fluid Dynamics	State	Coal Comb.	Size Distrib.	SO$_2$	NO$_x$	Re-Circulat
Mori	Block	Dyn	✔				✔
Basu	1.5-dim	Std	✔		✔		
Lin	1-dim	Std	✔		✔	✔	
Halder	1-dim	Std	✔				
IST	1-dim	Std	✔		✔	✔	
Alstrom	1-dim	Dyn	✔				
Haider	1.5-dim	Std	✔			✔	
IEA	1.5-dim	Std	✔	✔	✔	✔	✔

Std = steady state; Dyn = dynamic.

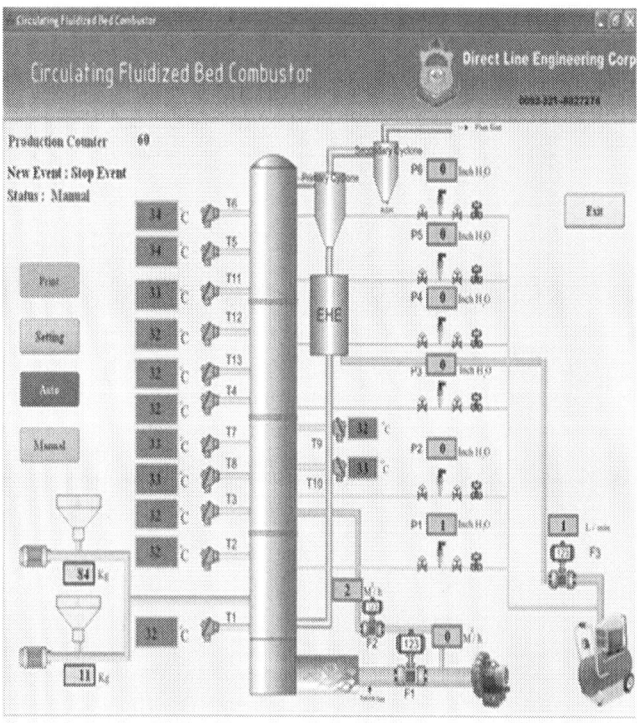

Figure 1: PLC view of CFB rig.

More detail regarding the experimental setup and operation can be seen elsewhere [4] . Silica sand, having sauter mean diameter (SMD) of 125 μm and particle density of 2500 kg/m^3 is used as the circulating bed material. Wheat straw (SMD = 0.85 mm) and Pakistani subbituminous coal (SMD = 0.49 mm) are used as the fuel in this study. Analyses and heating values of the feed materials are given in Table2 Reported values are the mean of three values taken as per ASTM standards. The concentrations of SO$_2$, NO$_x$ and CO in flue gas are measured by on line gas analyzers. Dry flue gas is also sampled in Teflon bags to analyze in gas chromatograph, Perkin Elmer Auto system GC Arnel. All reported values are corrected to 6% O$_2$ in the flue gas. Limestone (98.8% CaCO$_3$, SMD= 129 μm and ρ = 2730 kg/m^3) is also added as the sulfur capture sorbent through feeder.

Coal combustion model developed by Hannes [10] was used as a base model in this study. Effect of bed temperature, Ca/S molar ratio, solids circulation rate and secondary to primary air ratio, on the sulphur retention was predicted from the mathematical model at different blend ratios. Model results have been compared with the experimental results to see the reliability and synergy effect.

Modelling Approach

All main reactions were assumed to take place in the riser as in the return leg, temperature dropped and the availability of oxygen was small. For the use of matrix solver, it was reasonable to continue the annular phase into the dense bed, so that bed and freeboard could be solve together and continuously. The lateral mixing between core and annulus in the dense bed region was set high enough to equalize both phases to a common dense bed. All balances were setup by setting the time dependent term to zero to achieve steady state conditions. The gas flows were split using the values from the pre-calculations of the bubble holdup and the annulus width. The gas flow was balanced as molar flow. Changes caused by reactions which were not equimolar were assumed to have no influence on fluidization. The balanced flows were the convective flow in each phase (cor, ann, bub), cross flows from core to annulus (cor-ann), core to bubble (cor-bub) and vice versa, and mixing flows between the phases (corannx, corbubx).

Mass and Energy Balances

For mass balance, gaseous flows were balanced based on the following differential equation:

$$\frac{dn_g}{dt} = u_g \frac{\partial n_g}{\partial z} + \Psi_{source} + \Psi_{exchange} \tag{1}$$

An overall population balance was done to get the size distribution of the bed inventory. Then size classes (i) of the different materials (m), (coal, limestone, sand and ash) were balanced separately for each cell (L).

Table 2: Analysis of coal and wheat straw

	Proximate analysis			Elemental analysis					GCV
	VM	FC	Ash	C	H	N	S	O	MJ/kg
Salt range Coal (%)^{db}	38.60	43.90	17.50	68.90	9.8	0.56	4.2	16.54	25.55
Wheat Straw (%)^{db}	73.12	19.98	6.90	47.50	7.35	1.20	-	43.95	18.20

db = dry basis.

The following differential equation was discretized for each phase (cr, anl) considering size, materials and location (cell) of the solids.

$$\frac{dm_s}{dt} = u_s \frac{dm_s}{dz} + \Psi_{s,source}$$

$$\tag{2}$$

In the lowest cell of the riser, all annular material had to be returned to the core to conserve the mass balance.

Reactive species such as CaO and the combustibles in the coal were modelled as solid fractions. For better system solubility, the mass flow of the particles was kept constant, only the species fractions might vary. The fluidization pattern and char holdup was assumed not to be influenced by these changes. Coal mass was treated as virtual fraction, it did not influence the flow pattern but delivered the source terms for evaporated water and released volatiles. Only the ash residue in the char was balanced in the size distribution calculation. The fixed carbon was treated as ash load. The fraction balance was based on the following equation:

$$\frac{\partial(m_s \cdot x)}{\partial t} = u_s \frac{d(m_s \cdot x)}{dz} + \Psi_{s,source} + kM_s x \qquad (3)$$

Where the last term represents the release or reaction influence. "k" is a release or reaction constant depending on local gas concentrations and temperature. Drying and devolatilization are time dependent processes. The time dependent fraction was determined and averaged for each cell (L) and class (i) as under:

$$x_{avg}(i,L) = \int_0^\tau \frac{\partial x(i,L,t)}{\partial t} dt \qquad (4)$$

All these equations were written as first order equations for concentrations. The solution of the concentration equations is done analogue to the enthalpy balances.

The enthalpy balance delivered the average cell temperature. Enthalpy balance was based on the convective flows of gas and solids, changes in formation enthalpies due to reactions and the heat transfer to the walls. Following differential equation was used for the energy balance.

$$\left(n_g C_{pg} + m_s C_{ps}\right)\frac{dT}{dt} = \left(u_g n_g C_{pg} + u_s m_s C_{ps}\right)\frac{dT}{dz} + \Psi_{reac} + Q_{heatexch} \qquad (5)$$

The total balance system was consisted of the first order and solved through the matrix mechanism by arranging core and annular cells in the form of arrays. More detail about the sequence of calculations can be seen elsewhere [10] .

Sulphation Model

Lime stone is added into the CFB combustor to capture SO$_2$ directly. It is very tough to model the self-desulphurization of the coal, done by the mineral and metallic fractions in coal. The self-desulphurization of the coal is not explicitly modelled, however it can be taken into account by reducing the sulphur content in the coal by the amount of available calcium [25] . The capture of sulphur with limestone particles undergoes three principle reaction steps which are as follows:

calcination:

$$CaCO_3 \rightarrow CaO + CO_2 \tag{6}$$

oxidation:

$$SO_2 + 1/2O_2 \rightarrow SO_3 \tag{7}$$

sulphation:

$$CaO + SO_3 \rightarrow CaSO_4 \tag{8}$$

Overall reaction can be written as under:

$$CaCO_3 + 1/2O_2 + SO_2 \rightarrow CaSO_4 + CO_2 \tag{9}$$

Thus, the sulphur capture capability strongly depends on the residence time, the fragmentation behaviour, and the pore structure of the sorbent. Calcium-sulphur compounds do not only exist as $CaSO_4$ but may also exist as CaS, depending upon oxidizing or reducing boundary conditions, respectively. Since in fluidized beds the residence times of the particles are high and the sulphation reaction is slow, particle tracking is nearly impossible to distinguish in reducing and oxidizing zones as they mix within short times. Therefore, only the oxidizing conditions are considered in the following model.

During the calcination, the equilibrium between $CaCO_3$ and CaO is dependent on the partial pressure of CO_2 in the surrounding gas, and on temperature. Baker [26] stated, that this equilibrium pressure can be written as:

$$pCO_{2,eq} = 1.2 \times 10^7 \exp\left(-\frac{19124K}{T}\right)(bar) \tag{10}$$

For a given partial pressure of CO_2, calcination will only take place above the corresponding temperature. Dennis and Fieldes [27] have calculated the calcination time t_{calc} by:

$$t_{calc} = \frac{R_{p.0}M_{CaCO_3}}{k_0\left(pCO_{2.eq} - pCO_2 - pX_{CO_2}\right)} \tag{11}$$

With $k_0 = 207$ (mol/bar×m²×s) and an empirical variable describing a constant molar fraction of CO$_2$ which is 0.065 at 825°C, 0.1 at 875°C and 0.17 at 925°C.

For a kinetically controlled shrinking-core model, Kunii and Levenspiel [28] correlate the calcination time with the core radius and the conversion degree as:

$$\frac{t}{t_{calc}} = 1 - \frac{R_{core}}{R_{p,0}} = 1 - \left(1 - x_{calc}\right)^{0.33} \tag{12}$$

The combination with residence time of the particles is done with a residence time distribution:

$$E(t) = \frac{1}{\tau} \exp\left(-\frac{t}{\tau}\right) \tag{13}$$

Where the average residence time is the quotient of limestone mass in the furnace and limestone feed flow. The average calcination degree is

$$1 - x_{calc} = \int_0^{t_{calc}} \left(1 - \frac{t}{t_{calc}}\right)^3 \frac{\exp\left(-\dfrac{t}{\tau}\right)}{t} \, dt \tag{14}$$

On integration

$$x_{calc} = 3\left(\frac{\tau}{t_{calc}}\right) - 6\left(\frac{\tau}{t_{calc}}\right)^2 + 6\left(\frac{\tau}{t_{calc}}\right)^3 \left(1 - \exp\left(-\frac{t_{calc}}{\tau}\right)\right) \tag{15}$$

During calcination, the released CO$_2$ leaves the limestone with No. of pores which increases the inner surface area and subsequently sulphation reactions. Shrinking core model was used for the sulphation reaction due to its validity [24] as shrinking core models consider the "particle as a porous sphere", surrounded by a thin gas layer and consisting of an unreacted core in the particle surrounded by a shell of already sulphated material [28] . The radius of the unreached core shrinks with time enlarging the shell which causes a higher diffusion resistance for the penetrating gases [29] [30] .

Gas-solid reaction model, describing the reactions taking place at the individual particles is combined with the hydrodynamic model delivering the particle flow rates and concentrations. This model has differentials in time and radius, which are to be solved properly. So Wolff approach is implemented into the sulphation model, based on an analytical way to solve the radius dependent integral, so that only a forward integration in time remains [30]. The basic balance is the deliverance of the reactants by diffusion and the reaction at the surface of the unreacted core [23]:

$$4\pi r^2 \left(D_{SO_2} \cdot \frac{dC_{SO_2}}{dr} + D_{SO_3} \cdot \frac{dC_{SO_3}}{dr} \right) = 4\pi r \cdot k_{sulf} \cdot c_{SO_3} \quad (16)$$

Where the equilibrium between SO_2 and SO_3 can be expressed by:

$$c_{SO_3} = K_0 \cdot c_{SO_2} \sqrt{c_{O_2}} \quad \text{with } K_0 = 0.154 \sqrt{\frac{m^3}{mol}} \quad (17)$$

The reaction rate at the core surface can be stated as [10]:

$$\frac{dn_{CaO}}{dt} = -\gamma \cdot k_{sulf} \cdot 4\pi r^2 \cdot K_0 \cdot \sqrt{c_{O_2}} \cdot C_{SO_2 r} \quad (18)$$

Equation (18) can be solved as follows:

$$n_{CaO} = \frac{4}{3}\pi r^3 \cdot \frac{C_{O_2}\rho_{lime}}{M_{CaCO_3}} \cdot x_{calc} \cdot x_{CaCO_3} \quad (19)$$

Integration over the reacted shell and the gas film leads to the concentration of SO_2 on the core surface dependent on the bulk SO_2 concentration.

$$C_{SO_2 r} = \frac{C_{SO_2,R+\delta}}{1 + K_0 \sqrt{C_{O_2}} k_{sulf} r^2 \left(f_{film} + f_{shell} \right)} \quad (20)$$

Substitution back into Equation (16) and rewriting yields:

$$\frac{dr}{dt} = \frac{\gamma \cdot \dfrac{C_{SO_2,R+\delta}}{\dfrac{\rho_{lime}}{M_{CaCO_3}} \cdot x_{calc} \cdot x_{CaCO_3}}}{\dfrac{1}{k_s \cdot k_0 \cdot \sqrt{C_{O_2}}} + r^2 \left(f(D_{film}) + f(D_{shell}) \right)} \quad (21)$$

From the integration of left side of Equation (16) over the gas shell and over the reacted shell, the diffusion functions f_{film} and f_{shell} can be derived [25], which are:

$$f_{film} = \frac{1}{D_{SO_2,film} + D_{SO_3,film} K_0 \sqrt{C_{O_2}}} \left(\frac{1}{R} - \frac{1}{R+\delta} \right) \tag{22}$$

$$f_{shell} = \frac{1}{D_{SO_2,shell} + D_{SO_3,shell} K_0 \sqrt{C_{O_2}}} \left(\frac{1}{r} - \frac{1}{R} \right) \tag{23}$$

The conversion α_{lime} can be understood as reacted volume fraction of the particle

$$\alpha_{lime}(t) = 1 - \frac{r^3(t)}{R^3} \tag{24}$$

Averaged conversion is approached with a residence time distribution function [24] ,

$$\alpha_{lime} = \int_0^{\infty} \alpha_{lime}(t) \cdot E(t) dt \quad \text{where} \quad E(t) = \frac{1}{\tau} \cdot \exp\left(-\frac{t}{\tau} \right) \tag{25}$$

Above equation can be solved analytically using a substitution of Equation (24) into Equation (21).

$$\frac{1}{f(\alpha_{lime})} = \frac{1}{3\tau C_1} \left(C_2 (1-\alpha_{lime})^{\frac{2}{3}} + C_3 (1-\alpha_{lime})^{\frac{1}{3}} + (C_4 - C_3) \right) \tag{26}$$

With values of C_1, C_2, C_3 and C_4 as

$$C_1 = \gamma \frac{C_{SO_2,R+\delta}}{\dfrac{\rho_{lime}}{M_{CaCO_3}} x_{calc} x_{CaCO_3}}, \qquad C_2 = \frac{R}{k_{sulf} K_0 \sqrt{C_{O_2}}}$$

$$C_3 = \frac{R^2}{D_{SO_x,shell}}, \qquad C_4 = \frac{R^3}{D_{SO_x,film}} \left(\frac{1}{R} - \frac{1}{R+\delta} \right)$$

Final integration of Equation (26) is,

$$\int_0^{\alpha_{lime,max}} \frac{1}{f(\alpha_{lime})} = \frac{1}{\tau C_1} \left(C_2 \left(1 - (1-\alpha_{lime})^{\frac{1}{3}} \right) + \frac{C_3}{2} \left(1 - (1-\alpha_{lime})^{\frac{2}{3}} \right) + \frac{C_4 - C_3}{3} \alpha_{lime} \right) \tag{27}$$

This equation is replaced with residence time distribution function and numerically integrated using a modified Euler method. The function has as very steep gradient for very small values of α_{lime} and flattens with increasing values.

The diffusion coefficients consist of the Knudsen diffusion effects in the pores and the diffusion of a binary mixture of gases [21].

In the gas film, only binary diffusion occurs [22], so it can be assumed that $D_{SO_x,film} = D_{mix}$. In the shell, i.e. in the pores, Knudsen and gas diffusion must be considered [23]:

$$D_{SO_x,shell} = D_{pore} = \frac{\rho_{lime} \cdot V_{pore}}{\tau_{tort}} \cdot \left(\frac{1}{D_{knud}} + \frac{1}{D_{mix}} \right) \tag{28}$$

The calculation of the thickness of the gas film layer δ is estimated by the mass transport coefficient K_{SO_2} [14]:

$$\delta = \frac{1}{\dfrac{k_{SO_2,film}}{D_{SO_2,shell}} + \dfrac{1}{R}} \tag{29}$$

where

$$k_{SO_2} = \frac{St \cdot u_g}{\varepsilon_g}, \qquad St = 0.81 \cdot Re_p^{-0.5} \cdot Sc^{-0.66}$$

$$Sc = \frac{\mu_g}{\rho_g \cdot D_{mix}}, \qquad Re_p = \frac{\rho_g \cdot u_g \cdot d_p}{\mu_g}$$

So the reaction rates and concentrations were calculated and following parameters were used in the model $K_0 = 0.154$ (SO_4/SO_3 equilibrium constant) [35];

$t_{tort} = 3$ (tortuosity factor) [32];

$\alpha_{max} = 0.5$ (maximum conversion degree) [28];

$k_{sulf} = 0.15$ (sulphation constant m/s) [32].

Since the residence time of limestone particles and their sulphation takes place over hours, while gas residence time is in seconds, the sorbent is balanced as a homogenous phase. This is done by considering fragmentation and attrition of the sorbent which enlarges the available reactive surface. The conversion rate is calculated with an averaged SO$_2$ bulk concentration. The gas reaction is calculated depending on local holdup of sorbent in the riser. Weighing the local SO$_2$ concentrations with the local hold-up of sorbent provides the average gas concentration for the calculation to determine the conversion of the sorbent. The steady state sorbent conversion and gas concentration is established during the overall mass balance in the program.

To see the synergy effects and validate the model, its predictions were compared with the experimental data taken from the CFB test rig.

RESULTS AND DISCUSSIONS

Typical results obtained through the model and experimental studies are shown from Figures 2-6. There is a good agreement between the model predictions and the experimental results in accordance with the synergy effects of coal and biomass combustion, on emissions of SO$_2$. Model was run with a series of input values but reported values are, for bed temperature, excess air factor, secondary to primary air ratio, solid circulation rate and Ca/S molar ratio, for 5%, 10% and 20% blends of wheat straw with coal on weight basis.

It was believed that an increase in bed temperature can accelerate the calcination reaction resulting in low SO$_2$ concentration. High bed temperature also resulted in low CO concentrations which adversely affect the decompositions reactions of CaSO$_4$. Model predictions were in agreement when compared to the experimental values for different bed temperatures as shown in Figure 2. At higher temperature, higher conversion of lime stone and higher reaction rates of sulphation reactions were depicted in model as experimentally found to be happening in the furnace.

Agreement between model predictions and experimental results were found to be very encouraging for the effect of Ca/S molar ration on SO$_2$ emission as shown in Figure 3. Predicted values are more close to the experimental values for Ca/S molar ratios of 2 compared to that of 3. Model was also producing the reliable values at low wheat straw ratio in coal. Deviation at high wheat straw ratio in coal blend might be due to the different devolatilization kinetics of biomass compared to the coal.

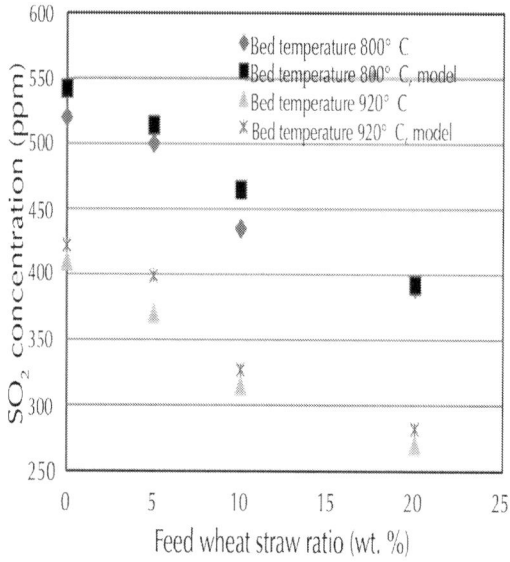

Figure 2: Bed temperature vs. SO_2 concentration, experimental results and model predictions.

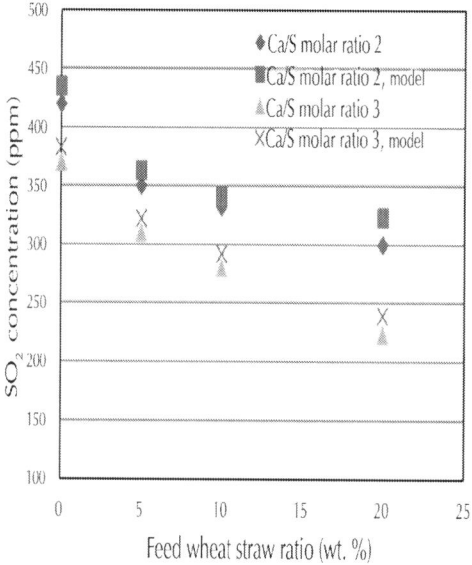

Figure 3: Ca/S molar ratio vs. SO_2 concentration, experimental results and model predictions.

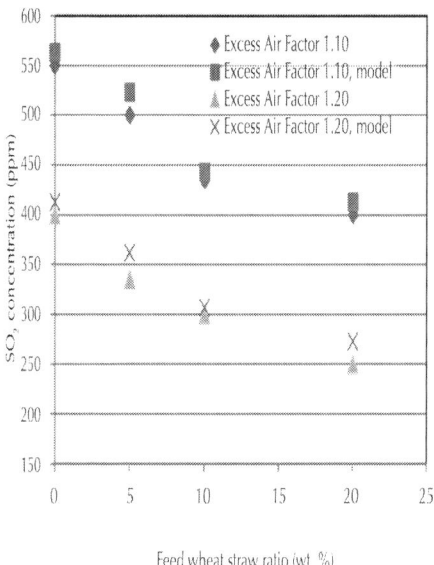

Figure 4: Excess air factor vs. SO$_2$ concentration, experimental results and model predictions.

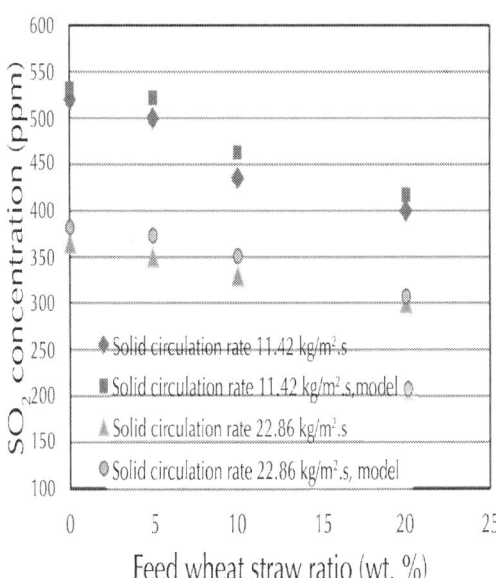

Figure 5: Solids circulation rate vs. SO$_2$ concentration, experimental results and model predictions.

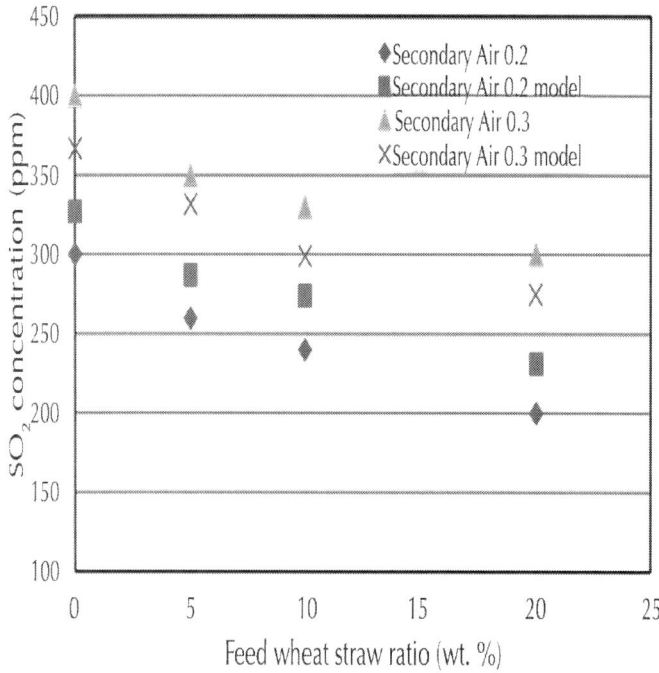

Figure 6: Secondary air to primary air ratio vs. SO$_2$ concentration, experimental results and model predictions.

Effect of variation of excess air factor on the SO$_2$ emission, predicted by the model is shown inFigure 4. Model has shown the correct tendencies for different values of excess air factor at different blends of wheat straw with coal on weight basis, although the model predictions were very slightly higher than the experimental ones. As incorporated in the model, increased concentrations of oxygen would facilitate the sulphation reaction, resulting in low SO$_2$ emission. With an excess air factor of 1.20, reaction rate of sulphation increased due to higher oxygen concentration in the riser. Based on the same scenario, SO$_2$ emission deceased in the actual experimental work.

Experimental and model results related to the effect of solids circulation rate on SO$_2$ emissions have been compared and reported in Figure 5. Model predictions have given the good relation in response to the variation in solid circulation rate, however minor deviations were also observed. At the value of 22.86 kg/m^2·s, the error was small and model has given the good predictions especially at higher wheat straw ratio.

In Figure 6, model predictions of effect of secondary to primary air ratio on SO$_2$ emission have been compared with the experimental results. As clear from the results, model was unable to produce good correlation for the variation in secondary to primary air ratio for the SO$_2$ emission. Model predictions have given a positive error for lower secondary to primary air ratio while a negative error was observed for the higher values of seconddary to primary air ratio. This might be due to the complex hydrodynamics inside the riser produced after the secondary air injection which could not be accounted in the present correlations used for the hydrodynamic modelling of the riser. As by the injection of secondary air, temperature of that region will be low and more oxidizing conditions would be made available making a precarious region regarding the model. Especially variation in the secondary to primary air ratio produced the undesired effect on the hydrodynamics of the riser that ultimately affected the SO$_2$ concentration.

CONCLUSION

A fluidized bed model for the steady state combustion and sulphation in a CFB was used to predict the SO$_2$ concentrations in the exit flue gases. It was based on the shrinking core model. Agreement between model prediction and experimental results was found encouraging for the parameters like bed temperature, fluidizing air velocity, excess air ratio and solids circulation rate. However for secondary to primary air ratio, some short comings in the model were observed.

ACKNOWLEDGEMENTS

Authors wish to thank the Higher Education Commission of Pakistan for their financial support to this project. (Project No. 1380-HEC).

REFERENCES

1.   Mitchell, C. and Connor, P. (2004) Renewable Energy Policy in the UK 1990-2003. Energy Policy, 32, 1935-1947. http://dx.doi.org/10.1016/j.enpol.2004.03.016

2.   Zhang, Y.N., Ghaly, A.E. and Li, B.X. (2012) Physical Properties of Wheat Straw Varieties Cultivated under Different Climatic and Soil Conditions in Three Continents. American Journal of Engineering and Applied Sciences, 5, 98-106.http://dx.doi.org/10.3844/ajeassp.2012.98.106

3.   Yang, S., He, H., Lu, S., Chen, D. and Zhu, J. (2008) Quantification of Crop Residue Burning in the Field and Its Influence on Ambient Air Quality in Suqian, China. Atmospheric Environment, 42, 1961-199.http://dx.doi.org/10.1016/j.atmosenv.2007.12.007

4.   Khurram, S., Mahmood, S., Waqar, A.K., Najaf, A. and Niaz, A.A. (2012) Parametric Study of NO_x Emissions in Circulating Fluidized Bed Combustor. Journal of Pakistan Institute of Chemical Engineers, 40, 61-68.

5.   McMullan, J. (2004) Fossil Fuel Power Generation. State-of-the-Art. Technical Report. Power Clean Thematic Network, Coleraine.

6.   Armesto, L., Boerrigter, H., Bahillo, A. and Otero, J. (2003) N_2O emissions from Fluidised Bed Combustion. The Effect of Fuel Characteristics and Operating Conditions [Small Star, Filled]. Fuel, 82, 1845-1850. http://dx.doi.org/10.1016/S0016-2361(03)00169-8

7.   Youssef, M.A., Wahid, S.S., Mohamed, M.A. and Askalany, A.A. (2009) Experimental Study on Egyptian Biomass Combustion in Circulating Fluidized Bed. Applied Energy, 86, 2644-2650. http://dx.doi.org/10.1016/j.apenergy.2009.04.021

8.   Sathitruangsak, P., Madhiyanon, T. and Soponronnarit, S. (2009) Rice Husk Co-Firing with Coal in a Short-Combustion-Chamber Fluidized-Bed Combustor (SFBC). Fuel, 88, 1394-1402. http://dx.doi.org/10.1016/j.fuel.2008.11.008

9.   Madhiyanon, T., Sathitruangsak, P. and Soponronnarit, S. (2009) Co-Combustion of Rice Husk with Coal in a Cyclonic Fluidized-Bed Combustor ([psi]-FBC). Fuel, 88, 132-138. http://dx.doi.org/10.1016/j.fuel.2008.08.008

10.   Hannes, J.P., Svoboda, K. and van den Bleek, C.M. (1995) The IEA Model for Circulating Fluidized Bed Combustion. 13th International Conference on Fluidized Bed Combustion, Orlando, 287-296.

11. Hartleben, B. (1983) Mathematische modellierung von blasenbildenden kohlewirbelschicht-feuerungsanlagen. Ph.D. Thesis, Siegen University, Siegen.

12. Zhang, W., Tung, Y. and Johnsson, F. (1991) Radial Voidage Profiles in Fast Fluidized Beds of Different Diameters. Chemical Engineering Science, 46, 3045-3052.http://dx.doi. org/10.1016/0009-2509(91)85008-L

13. Mori, S., Narukawa, K., Yamada, I. and Tanii, H. (1991) Dynamic Model of a Circulating Fluidized Bed Coal Fired Boiler. 11th International Conference on Fluidized Bed Combustion, Montreal, 21-24 April 1991, 1261-1265.

14. Mori, S. and Wen, C.Y. (1975) Estimation of Bubble Diameter in Gaseous Fluidized Beds. AlChE Journal, 21, 109-115. http:// dx.doi.org/10.1002/aic.690210114

15. Sengupta, S.P. and Basu, P. (1991) A Generalized Mathematical Model for Circulating Fluidized Bed Boiler Furnace. 11th International Conference on Fluidized Bed Combustion, Montreal, 21-24 April 1991, 1295-1301.

16. Talukdar, J., Basu, P. and Joos, E. (1993) Sensitivity Analysis of a Performance Predictive Model of Circulating Fluidized Boiler Furnace. 4th International Conference on Circulating Fluidized Beds, Somerset, 1-5 August 1993, 541-546.

17. Xu, X. and Mao, J. (1993) Mathematical Model and Simulation of Circulating Fluidized Bed Boilers. 4th International Conference on Circulating Fluidized Beds, Somerset, 1-5 August 1993, 104-109.

18. Lin, X. and Li, Y. (1993) A Two Phase Model for Fast Fluidized Bed Combustion. 4th International Conference on Circulating Fluidized Beds, Somerset, 1-5 August 1993, 547-552.

19. Halder, P.K. and Datta, A. (1993) Modelling of Combustion of a Char in a Circulating Fluidized Bed. 4th International Conference on Circulating Fluidized Beds, Somerset, 1-5 August 1993, 92-97.

20. Saraiva, P.C., Azevedo, J.L.T. and Carvalho, M.G. (1993) Modelling the Flow, Combustion and Pollutants Emission in a Semi Industrial CAFBC. 4th International Conference on Circulating Fluidized Beds, Somerset, 1-5 August 1993, 72-79.

21. Haider, A. and Levenspiel, O. (1989) Drag Coefficient and Terminal Velocity of Spherical and Nonspherical Particles. Powder Technology, 58, 63-70. http://dx.doi.org/10.1016/0032-5910(89)80008-7

22. Hiller, R. (1995) Mathematische modellierung der kohleverbrennung in einer circofluid wirbelschichtfeuerung. Ph.D. Thesis, Dortmund University, Dortmund.

23. Hannes, J.P., Svoboda, K. and van den Bleek, C.M. (1993) Mathematical Modelling of CFBC: An Overall Modular Programming Frame Using a 1.5-Dimensional Riser Model. 12th International Conference on Fluidized Bed Combustion, San Diegi, 9-13 May 1993, 455-463.

24. Hannes, J.P., Svoboda, K. and van den Bleek, C.M. (1993) Modelling of Size Distribution and Pressure Profiles in a CFBC. 4th International Conference on Circulating Fluidized Beds, Somerset, 1-5 August 1993, 199-204.

25. Hannes, J.P. (1996) Mathematical Modelling of Circulating Fluidized Bed Combustion. Ph.D. Thesis, Technical University Delft, Delft.

26. Baker, E.H. (1962) the Calcium Oxide-Calcium Dioxide System in the Pressure Range 1 - 300 Atmospheres. Journal of Chemical Society, 165, 464-470.http://dx.doi.org/10.1039/jr9620000464

27. Dennis, J.S. and Fieldes, R.B. (1986) Simplified Model for the Rate of Sulphation of Limestone Particles. Chemical Engineering Research and Design, 64, 279-287.

28. Kunii, D. and Levenspiel, O. (1991) Fluidization Engineering. 2nd Edition, Butterworth-Heinemann, Boston.

29. Korbee, R. (1995) Regenerative Desulphurization in an Interconnected Fluidized Bed System. Ph.D. Thesis, Technical University Delft, Delft.

30. Wolff, E.H.P. (1991) Regenerative Sulphur Capture in Fluidized Bed Combustion of Coal. Ph.D. Thesis, Technical University Delft, Delft.

31. Hartman, M. and Trnka, O. (1993) Reactions between Calcium-Oxide and Flue Gas Containing Sulfur-Dioxide at Lower Temperatures. AIChE Journal, 39, 615-624. http://dx.doi.org/10.1002/aic.690390410

32. Dam, J. and Ostergarrd, K. (1991) High Temperature Reaction between Sulphur Dioxide and Limestone in Two Reactors. Chemical Engineering Science, 46, 822-831.

33. Hartman, M., Svoboda, K., Trnka, O. and Vesely, V. (1999) Removal of Sulphur from Hot Coal Gas. Chemicke Listy, 93, 99-106.

34. Hartman, M., Svoboda, K., Trnka, O. and Vesely, V. (1988) Reaction of Sulfur-Dioxide with Magnesia in a Fluidized-Bed. Chemical Engineering Science, 43, 2045-2050. http://dx.doi.org/10.1016/0009-2509(88)87082-9

35. Hansen, F.B.P., Lin, W. and Dam-Johansen, K. (1997) Chemical Reaction Conditions in a Danish 80 MWth CFB-Boiler Co-Firing Straw and Coal in Fluidized Beds. In: Preto, F.D.S., Ed., 14th International Conference on Fluidized Bed Combustion, ASME, Vancouver & New York, 287-294.

Dehydrocyclization of n-Hexane Over Heteropolyoxometalates Catalysts

Abdellah Eid[1], Ouarda Benlounes[1], Hikmat S. Hilal[2], Chérifa Rabia[3], and Smain Hocine[1]

[1]Laboratoire de Chimie Appliquée et de Genie Chimique, Universite Mouloud Mammeri, Tizi-Ouzou, Algeria

[2]Laboratory of Semiconductor and Solar Energy Research, Chemistry Department, An-Najah N. University, Nablus, Palestine

[3]Laboratoire de Chimie du Gaz Naturel, Institut de Chimie USTHB, Alger, Algeria

ABSTRACT

The catalytic dehydrocyclization of n-hexane was studied here for the first time using a number of compounds based on $H_3PMo_{12}O_{40}$. The described catalysts were prepared by either replacing the acidic proton with counter-ions such as ammonium or transition metal cations (NH_4^+, Fe^{3+}, K^+), or by replacing Mo^{6+} with (Ni^{3+}, Co^{3+}, Mn^{3+}) in the polyoxometalate framework, as reported earlier. For comparison purposes, the known (TBA) $_7PW_{11}O_{39}$ catalyst system was used. All reactions were conducted at different temperatures in the range 200°C - 450°C. The Keggin structure of these heteropolycompounds was ascertained by XRD, UV and IR measurements. ^{31}P NMR measurements and thermal behaviour of the prepared catalysts were also studied. These modified polyoxometalates exhibited heterogeneous superacidic catalytic activities in dehydrocyclization of n-hexane into benzene, cyclohexane, cyclohexene and cyclohexadiene. The catalysts obtained by substituting the acidic proton or coordination atom exhibited higher selectivity and stability than the parent compound $H_3PMo_{12}O_{40}$. Catalytic activity and selectivity were heavily dependent on the composition of the catalyst and on the reaction conditions. At higher temperatures, the catalyst exhibited higher conversion efficiency at the expense of selectivity. Using higher temperatures (>400°C) in the presence of hydrogen carrier gas, selectivity towards dehydrocyclization ceased and methane dominated. To explain the results, a plausible mechanism is presented, based on super-acidic nature of the catalyst systems.

INTRODUCTION

Catalytic reforming of n-hexane is one of the most widely used processes in oil refining industry. Fuel octane number can be enhanced by increasing concentrations of aromatic compounds [1-3].

Catalytic reforming proceeds over bifunctional catalysts having both metallic and acidic functions are currently being considered [4, 5]. Hetero poly acids (HPAs) are complex proton acids that incorporate polyoxometalate anions (hetero poly anions) having metal-oxygen octahedra as the basic structural units. HPAs have several advantages such as very strong Brönsted acidities approaching the super acid region [6] and high oxidant activities in multielectron redox reactions under mild conditions. Their acid-base and redox properties can be varied widely by changing their chemical compositions.

HPAs have a discrete ionic structure, containing the fairly mobile hetero poly anions and counter-cations (H^+, H_3O^+, $H_5O_2^+$, etc.). This unique structure results in significantly high proton mobility and pseudo liquid phase behaviour. In addition, HPAs are highly soluble in polar solvents and their thermal stabilities are quite high. These properties make HPAs potentially promising acid, redox, and bifunctional catalysts in both homogeneous and heterogeneous systems [7, 8].

Hetero poly acids (HPA) usually act as strong acidic catalysts, with the ability to activate alkanes and produce carbonium ion intermediates [9, 10]. Such intermediates are necessary for alkane cracking, alkylation and isomerization processes [11]. The large-scale industrial applications of these reactions drew attention to such catalyst systems for long time [12].

The transformation of n-hexane depends on the nature of the catalyst and on the operation conditions. The reactant may undergo three different competing reactions, namely: Hydrogenolysis to short linear chain alkanes such as methane, ethane, propane, butane and pentane in presence of H-[Al]-ZMS-5 leading [13]. Isomerization over Pt/zeolithe giving iso-alkanes, such as isobutane, isopentane, 2, 2-dimethylbutane, 2, 3-dimethylbutane, 2- methylpentane and 3-methylpentane [14] Dehydrocyclization and aromatization giving cyclohexane, cyclohexene, and benzene on catalysts such as WO_3/ZrO_2 [15].

In our search for highly efficient, stable and selective catalyst systems for dehydrocyclization of hexane, synthesised and characterised a number of compounds based on modified polyoxomolybdates and polyoxytengestates. The investigated compounds are the heteropolyacid $H_3PMo_{12}O_{40} \cdot xH_2O$ (noted PMo_{12}), the ammonium salt of the (Co, Ni, Mn) substituted species $(NH_4)_4PMo_{11}O_{39}M (H_2O) \cdot xH_2O$ (M: Co, Ni, Mn) (noted $PMo_{11}Co$, $PMo_{11}Ni$, $PMo_{11}Mn$) $(TBA)_7PW_{11}O_{39} \cdot xH_2O$ (noted PW_{11}) and $KFePMo_{12}O_{40} \cdot xH_2O$ (noted $FePMo_{12}$). Such complexes have been intensively studied for different organic reactions [16-20], and can activate alkanes due to their superacidic nature [21-24].

In this work, the compounds have been used in hexane dehydrocyclization reaction for the first time. The reaction may involve other products such as cyclohexane, cyclohexene and cyclohexadiene, in addition to the unwanted methane product. The main goal is to maximize benzene production and minimize methane in the product mixture, by investigating different catalysts, varying temperature, and carrier gas and retention time.

EXPERIMENTAL PART

Preparation of Catalysts

Pure heteropolyacid $H_3PMo_{12}O_{40} \cdot xH_2O$ was prepared as described earlier [25]. The mixed potassium-iron salt $K_{0.5}Fe_{0.1}H_{0.2}PMo_{12}O_{40} \cdot xH_2O$ was prepared as follows: to an aqueous solution of $H_3PMo_{12}O_{40}$ (50 ml, 0.08 M) were added successively 9.0 ml of 0.10 M Fe $(NO_3)_3$ and 7.5 ml of 0.80 M KNO_3. The obtained precipitate was filtered over glass frit and dried at 50°C under vacuum for 5 h.

The ammonium salt $(NH_4)_4PMo_{11}MO_{39}$ $(H_2O) \cdot xH_2O$ (M = Ni, Co, Mn) was prepared as described earlier [26]. A mixture of H_3PO_4 (5.0 ml; 1.00 M), H_2SO_4 (10.0 ml; 0.50 M) and MSO_4 (5.0 ml; 1.00 M) was slowly added to 250 ml of an aqueous solution of ammonium paramolybdate $[(NH_4)_6Mo_7O_{24} \cdot 4H_2O]$ (48.8 g, 40 mmol) at 0°C. The ammonium salt completely precipitated after addition of solid ammonium nitrate (NH_4NO_3).

(TBA) $_7PW_{11}O_{39}$ was prepared by the method described in the literature [27]. A 15.00 g quantity of $H_3PW_{12}O_{40} \cdot 14H_2O$ was dissolved in 20.0 mL of water; a 5.0 g quantity of $[(n-C_4H_9)_4N]$ Br (TBABr) was dissolved in 50.0 mL of CH_2Cl_2. The mixture was poured into the molybdophosphate solution with vigorous stirring. Three phases were formed: a yellow solid one, a lower yellow liquid one (CH_2Cl_2), and a poorly yellow coloured aqueous upper one. The solid was washed with Et_2O, providing 8.50 g of product. A further yield of 2.50 g was obtained by addition of Et_2O to the CH_2Cl_2 phase. Recrystalization was performed in CH_2Cl_2.

Equipment

The UV analysis of the catalyst samples (dissolved in acetonitrile/water 1:1 ratio, as dilute samples solutions 10-5-10-3 M) were performed on a UV-1601PC SHIMADZU spectrophotometer. IR spectra of the samples were obtained at room temperature with a BIO-RAD FTS 165 FTIR spectrometer using standard KBr pellet technique. Measurements were taken in the wave-number range of 4000 - 400 cm^{-1}, 256 scans, and resolution 2 cm^{-1}.

The X-ray powder diffraction (XRD) diagrams were recorded on a SIEMENS D5000 equipment in the 5 - 45° 2θ range, at a scanning speed of 1°C per minute, using CuK$_\alpha$ radiation. The ^{31}P NMR measurements were performed on a Bruker CXP10 spectrophotometer at 10 MHz. Chemical shifts were referenced to Al (PO$_3$)$_3$.

Thermal behaviour of the prepared catalysts was studied using a tgA-dta Setaram apparatus. For each experiment 20 mg of powdered sample were heated at a temperature ramp rate of 5°C/min under air flow. Differential scanning calorimetry experiments (DSC) were run on a DSC 111 SETARAM. Apparatus under similar conditions, with a nitrogen flow.

Catalytic Experiments

The conversion of n-hexane was carried out in a quartz fixed-bed flow reactor at different temperatures from 200°C to 400°C, using 0.5 g of catalyst. Catalysts were heated to the desired reaction temperature at 5°C/min under nitrogen flow and then maintained at this temperature for one hour under nitrogen. A mixture of n-hexane (50 torr) and hydrogen was flowed over the catalyst (0.1 l/h). The reaction mixture was analysed by gas chromatography on a Hewlett Packard 5830A apparatus equipped with two successive separation columns, filled respectively with ®Porapak QS for a TCD detector and 5Å-molecular sieve for FIDCG detector.

Reaction rates (r) were calculated from the following equation:

$$r\left(mol\cdot h^{-1}\cdot g^{-1}\right) = F/22.4\times 273/T \times p/760 \times 1/m \tag{1}$$

Where F is the flow rate of the vector gas (l/h), T the reaction temperature (K), p the partial pressure of the products (torr) and m the mass of the catalyst (g).

The activation energy (E$_a$) was obtained from the Arrhenius plot according to:

$$r = k \times \exp\left(-E_a/RT\right) \tag{2}$$

Where R is the perfect gas constant and k the pre-exponential factor.

RESULTS AND DISCUSSION

Characterization of Catalysts

In the UV, the Keggin heteropolyanions show two absorption bands, one around 200 nm and another around 280 nm. According to literature [28], when catalyst concentration inside the suspension decreases, the two bands disappear gradually indicating decomposition of the polyanion with dilution [29]. Therefore, intermediate concentrations of different catalyst systems were used here for characterization, as shown in Figure 1.

Values of absorption bands are summarized in Table 1. The Table shows main band observed for each prepared catalyst. The bands are assigned respectively to the vibrations of terminal Mo = Ot and bridging bonds. The bridging bonds are two types: the inter-bridges (M-Ob-M) that occur between two adjacent octahedra, and the intrabridges (M-Oc-M) that occur within same octahedron.

Solid-state FT-IR spectra measured for different prepared catalysts are given in Figure 2. The spectra display similar patterns characteristic for the Keggin structure [30, 31] Values of IF bands are summarized in Table 2.

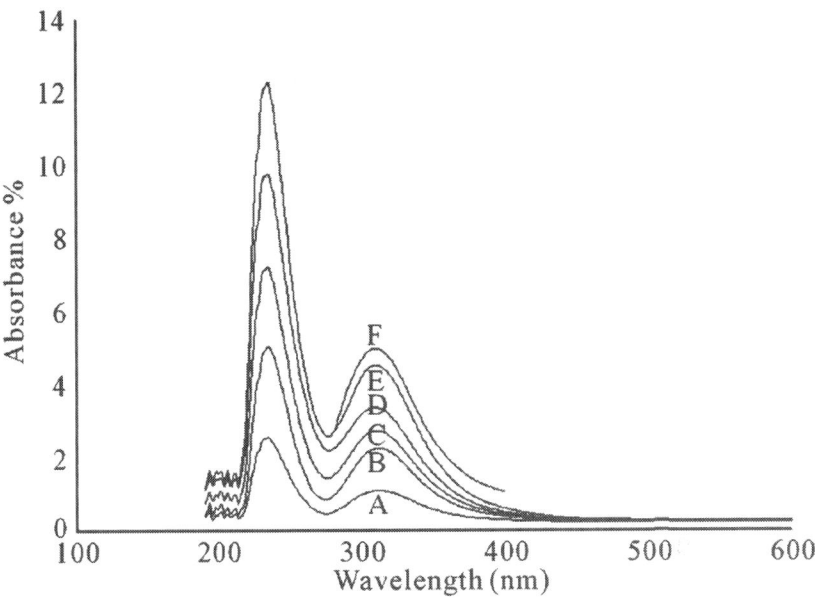

Figure 1: UV spectra for different catalyst systems (A: PMo_{12}; B: $FePMo_{12}$; C: $PMo_{11}Ni$; D: $PMo_{11}Co$; E: $PMo_{11}Mn$; F: PW_{11}). A suspension of each catalyst system in acetonitrile/water mixture was used for UV absorption spectra.

Table 1: Main UV bands observed for different HPA catalyst systems

Catalyst	λ_{max} (nm)	
	M = Ot	**M-Ob**
PMo_{12}	237	310
PW_{11}	248	311
$PMo_{11}Co$	232	306
$PMo_{11}Ni$	236	306
$PMo_{11}Mn$	237	311
$KFeMo_{12}O_{40}$	235	306

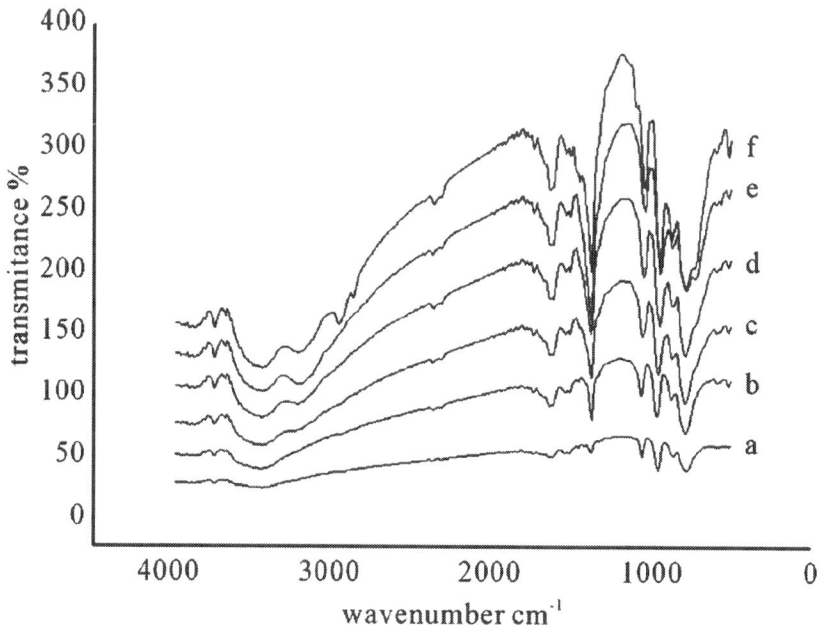

Figure 2: Solid state FT-IR spectra measured as KBr discs for a: PW_{11}; b: $PMo_{11}Mn$; c: $PMo_{11}Ni$; d: $PMo_{11}Co$; e: $KFePMo_{12}$; f: PMo_{12}.

Table 2: FT-IR absorption band values for different catalyst systems, together with their assignments

Catalys	$P-O_a$	$Mo-O_t$	$Mo-O_b-Mo$	$Mo-O_c-Mo$
$HPMo_{12}$	1064	962	868	789
$PMo_{11}Ni$	1047	945	870	725
$PMo_{11}Mn$	1108	933	900	725
$PMo_{11}Co$	1047	943	806	737
PW_{11}	1059	962	884	797
$KFePMo_{12}$	1064	962	868	789

FT-IR spectra show that main peaks around 1064, 961, 861 and 786 cm^{-1} are attributed to $P-O_a$, $Mo = O_t$, MoO_b-Mo, and $Mo-O_c$-Mo bonds, respectively.

^{31}P NMR spectra were measured for only two catalyst types, as shown in Figures 3(a) and (b).Figure 3(a) shows a single peak at −3.47

ppm for salt $Fe_{0.1}PMo_{12}O_{40}$, which indicates purity of the sample. The KPW_{11} heteropolyanion showed a signal at 10.35 ppm, Figure 3(b).

The signal is characteristic for the lacunary species $K_7[\alpha\text{-}PW_{11}O_{39}]$ [32, 33]. ^{31}P NMR measured and literature data are summarized in Table 3.

XRD patterns were measured for all solid catalysts systems in the powder form. Figure 4 shows the diffractograms of $HPMo_{12}$, $FePMo_{12}$ and $PMo_{11}Ni$. The diffractogram of $H_3PMo_{12}O_{40}\cdot\textbf{13H}_2\textbf{O}$ was consistent with literature [36] and corresponded to a triclinic structure.

Other diffractograms show that the salt $FePMo_{12}$ crystallises in a cubic structure with intense (222) peak [16, 37]. This corresponds to same diffraction plane for ammonium salts of ($PMo_{11}Ni$, $PMo_{11}Co$, and $PMo_{11}Mn$) with a monoclinic structure.

Thermogravimetric (TGA) analysis of the $H_3PMo_{12}O_{40}$ acid shows two mass loss signals between 40°C and 140°C corresponding to the departure of water of crystallization or hydration. A second mass loss at temperatures between 250°C and 350°C was attributed to water content resulting from combination of H^+ ions and network oxygen. This mass loss leads to a reversible modification of the polyanion.

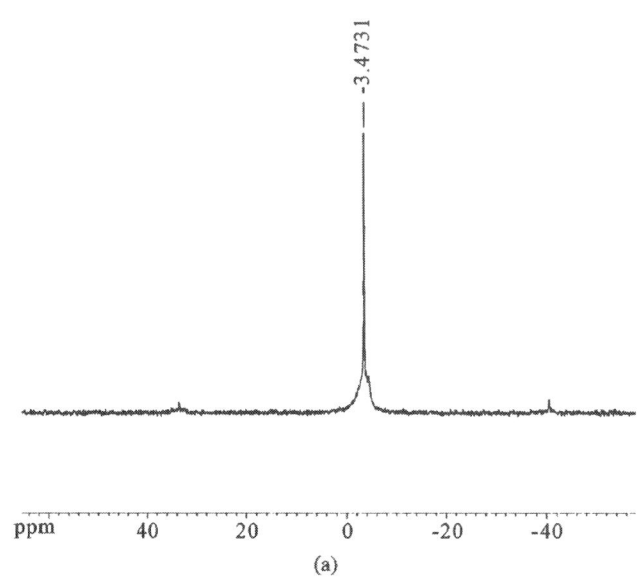

-3.4731

| ppm | 40 | 20 | 0 | -20 | -40 |

(a)

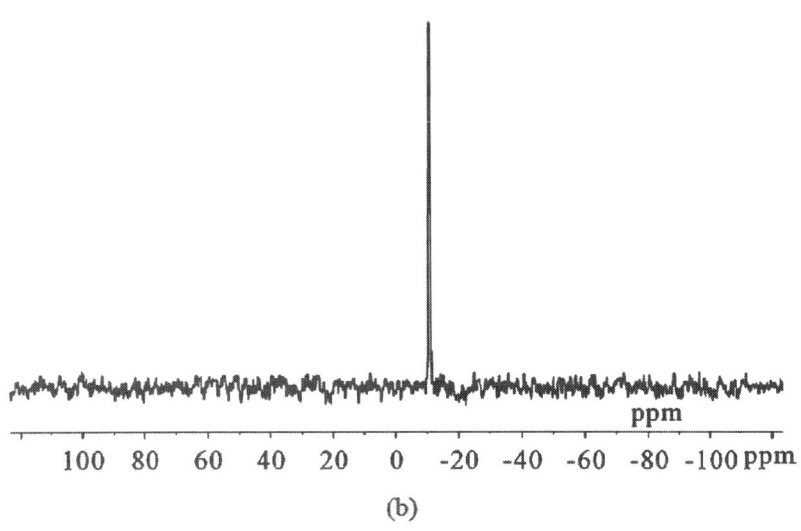

ppm

100   80   60   40   20   0   -20 -40 -60 -80 -100 ppm

(b)

Figure 3: ^{31}P NMR spectra for: (a) FePMo$_{12}$; (b) PW$_{11}$.

Table 3: ^{31}P NMR measured and literature data for different catalyst systems

No.	Catalysts	Chemical shift ppm	Ref.
1	PMo$_{12}$	−3.7	[34]
2	PMo$_{12}$Ni	−1.1551	[35]
3	PMo$_{11}$Co	−4826	[35]
4	PW$_{11}$	−10.35	This work
5	FePMo$_{12}$	−3.47	This work

In differential thermal (DT) analysis, the loss of water corresponds to endothermic peaks. An exothermic peak was observed above 350°C and was attributed to acid decomposition into P$_2$O$_5$ and MoO$_3$ oxides.

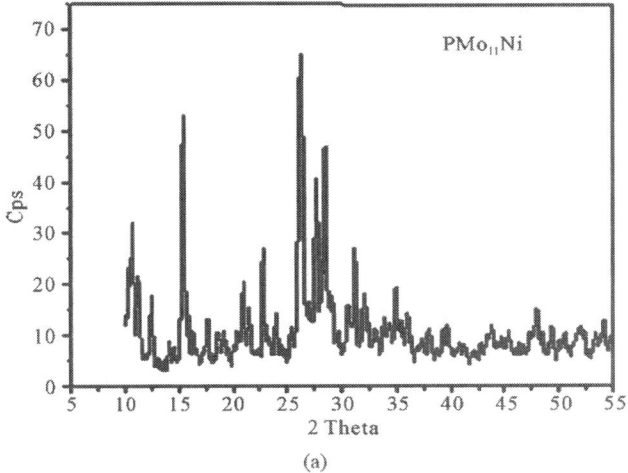

(a)

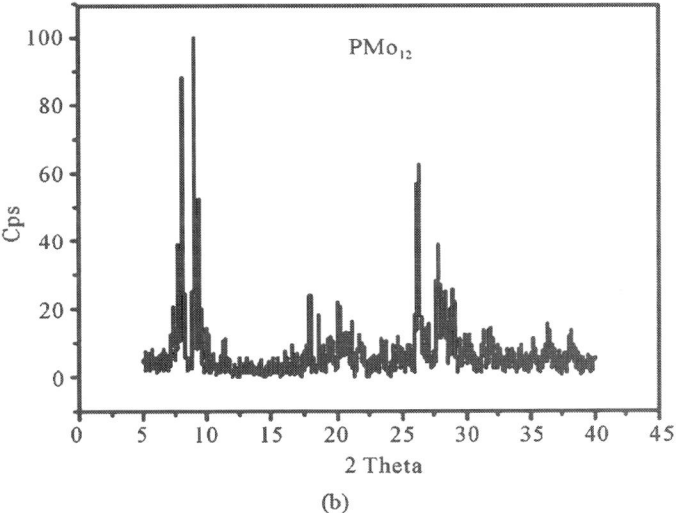

(b)

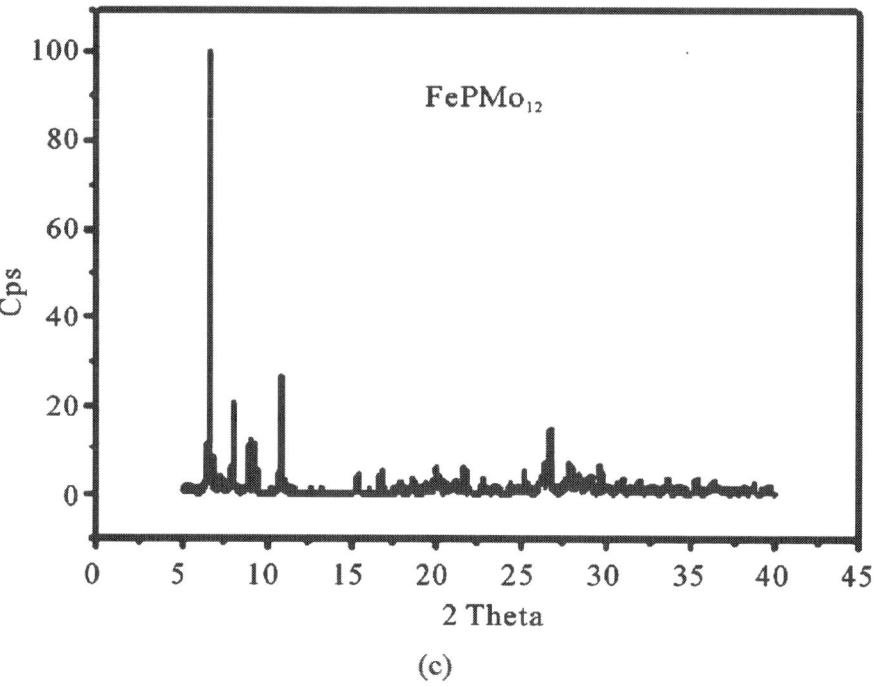

(c)

Figure 4: XRD spectrum of: (a) $PMo_{11}Ni$; (b) PMo_{12}; (c) $FePMo_{12}$.

TGA analysis for the salts $PMo_{11}Ni$, $PMo_{11}Co$, $PMo_{11}Mn$, $FePMo_{12}$, and PW_{11} shows three mass losses. A first mass loss occurred around 100°C, which is interpreted as the start of physisorbed water loss. A second departure is at around 240°C to 300°C, which is due to loss of constitution water molecule. Water molecules coordinated to nickel and cobalt ions are involved here. Finally, a loss around 300°C to 400°C, corresponding to departure of molecular ammonia or molecular nitrogen, was observed. Salt catalysts exhibit higher thermal stability than $H_3PMo_{12}O_{40}$ (Figure 5). The temperature at which their degradation starts depends on the position of the ion M^{3+}. When the counter ion involved metal ionic species, the decomposition temperature exceeded 500°C. On the other hand, the heteropolyacid $H_3PMo_{12}O_{40}$ is less thermally stable than the salt, and its decomposition started at 350°C.

Catalytic Study

Effect of Catalyst Type

Steady-state activity for catalytic n-hexane dehydrocyclization was reached within about 4 h. The observed products were benzene, (one most preferred product), cyclohexane, cyclohexene, cycloheadiene and methane (least preferred product). As expected, the catalytic activity and selectivity were dependent on the nature of the catalyst.

Figure 6 shows the conversion of n-hexane using different catalysts. The conversion of the solid catalyst was sensitive to the nature of the coordination ion (Mo^{6+}, W^{6+}, Ni^{3+}, Co^{3+}, Mn^{3+}) and counter ion H_3O^+, NH_4^+, Fe^{3+}). The replacement of mobile M (Ni^{3+}, Co^{3+}, Mn^{3+}) ions in the HPA framework increased the activity of the catalyst. Replacing H^+ ion with Fe^{3+} lowered the activity. PW_{11} exhibited relatively high activity as well.

Detailed studies on n-hexane isomerization catalysis with HPC are summarized in (Tables 5 and 6). The results show that benzene was the major product at low conversion level while using different catalysts. Cyclohexane, cyclohexene, cyclohexadiene, benzene and methane were also observed. As the conversion level increased benzene product ratio was lowered in ratio of methane product.

The 30% activity for $H_3PMo_{12}O_{40}$ (or even 100% for other catalysts excluding FePMo) shows that the protons in anhydrous HPA are accessible to the reactant molecules. The high catalytic activity of HPA may thus be related to high acid strength and high mobility of protons, as reported earlier [38]. The resulting protons are responsible of the aromatization of n-hexane. Cyclohexane, cyclohexene cyclohexadiene were also obtained, with variable ratios depending on type of catalyst. An important selectivity obtained by PMoNi catalyst involved cyclohexane 14%, cyclohexene 2%, cyclohexadiene 40%. A lower selectivity with PMoCo and PMoMn catalysts was observed. The results indicate that both relatively strong acidity and metallic propriety are responsible for catalyst efficiency at lower temperature (lower than 350°C).

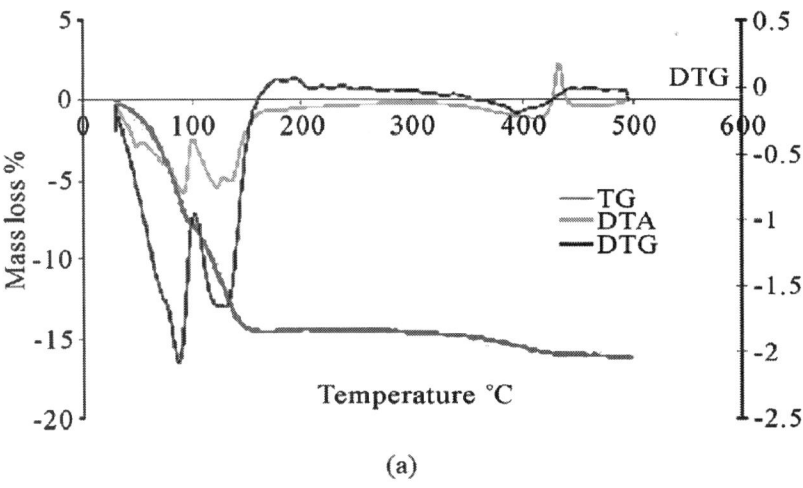

(a)

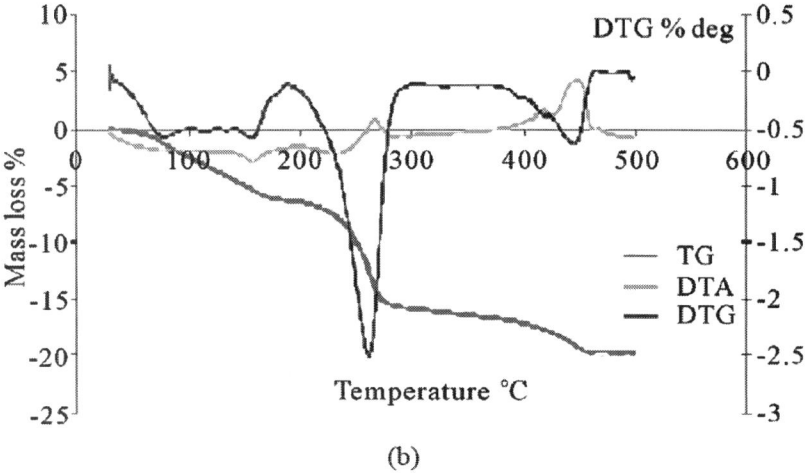

(b)

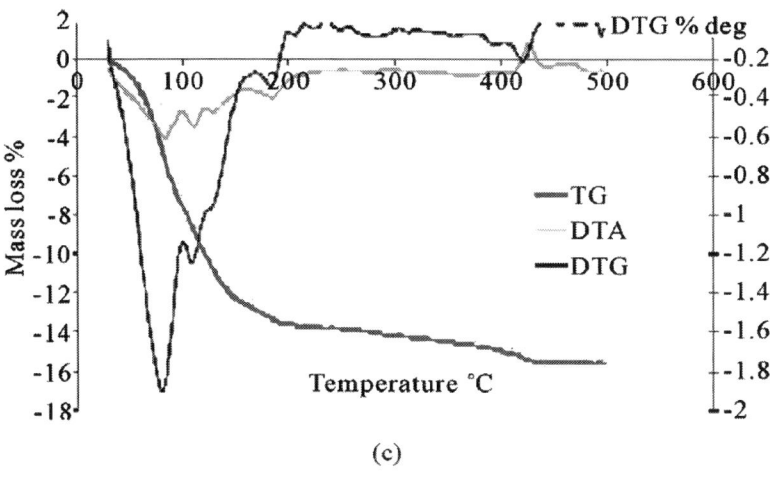

(c)

Figure 5: Thermal analysis of: (a) PMo_{12}; (b) $PMo_{11}Ni$; (c) $FePMo_{12}$.

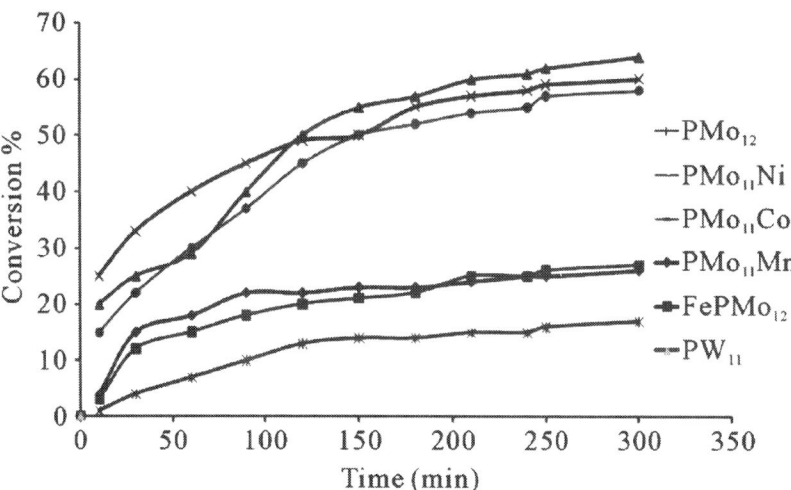

Figure 6: Conversion of n-hexane as function of time at 350°C, (using N_2 as carrier gas).

Table 4: Lattice parameters (A°) measured for different catalyst systems

Heteropolycompounds	Parameters (A°)		Ref.
$H_3PMo_{12}O_{40} \cdot 13H_2O$	a = 14.10	$\alpha =$ 112.1°	[35,36]
	b = 14.13	$\beta =$ 109.8°	
	c = 13.39	$\gamma = 60.73°$	
	triclinic		
$(NH_4)_4HPMo_{11}NiO_{39}$	a = 20.072	b = 11.55938	[35]
	c = 18.77828	$\beta =$ 109.58°	
	monoclinic		
$(NH_4)_4HPMo_{11}CoO_{39}$	a = 21.01022	b = 7.8468	[35]
	c = 17.00056	$\beta =$ 97,539°	
	monoclinic		
$(NH_4)_4PMo_{11}MnO_{39}$	a = 21.12322	b = 7.19688	[35]
	c = 17.00056	$\beta =$ 97.653°	
	monoclinic		

Table 5: Catalytic performances of catalysts (Carrier gas H_2, flow rate: 0.1 l\h)

Catalyst	T °C	Con. %	Selectivity				
			CH_4	C_6H_6	C_6H_{12}	C_6H_{10}	C_6H_8
	300	5	79	21			
PMo_{12}	350	30	83	17			
	380	100	100				
	250	48	53	21	7	12	7
	300	77	56	21	8	9	6
$PMo_{11}Ni$	350	100	85	12	3		
	380	100	100				
	250	35	57	23	20		
	300	50	86	8	6		
$PMo_{11}Co$	350	100	98		2		
	380	100	100				
	250	60	90	7	3		
	300	80	98		2		
$PMo_{11}Mn$	350	100	100				
	380	100	100				
	250	56	52	20		12	12
	300	90	89	11			
$Fe_{0.1}PMo_{12}$	350	100	100				
	380	100	100				
PW_{11}	300	3	0	100			
	350	69	0	100			

Table 6: Catalytic performances of catalysts (carrier gas N_2, flow rate: 0.1 L\h)

Catalyst	T °C	Con. %	Selectivity				
			CH_4	C_6H_6	C_6H_{12}	C_6H_{10}	C_6H_8
PMo_{12}	300	2	0	100			
	350	23	20	80			
	380	100	60	40			
$PMo_{11}Ni$	250	1	0	78	12	2	8
	300	8	0	52	24	4	20
	350	24	7	37	14	2	40
	400	100	100	0			
$PMo_{11}Co$	250	4	0	100			
	300	11	0	100			
	350	60	58	42			
	400	100	100	0			
$PMo_{11}Mn$	250	3	100	0			
	300	11	0	100			
	350	58	53	47			
	400	100	100	0			
$Fe_{0.1}PMo_{12}$	300	6	0	100			
	350	15	20	80			
	380	50	80	20			
	400	86	100	0			
PW_{11}	300	3	0	100			
	350	57	0	100			
	400	67	0	100			

The PW_{11} catalyst showed especially high efficiency, selectivity and stability towards benzene formation. For example, PW_{11} showed high activity (69%) conversion and high selectivity (100%) at 350°C compared to PMo_{12} (30% conversion and 17% selectivity), PMoNi (100% and 11.5%), and PMoCo and PMoMn (100% and 0%). Such behaviour should be ascribed to the Lewis acid properties of the TBA cations; also, we may conclude that Lewis acid species should be important components of the active site in the isomerization of n-hexane reaction over HPAs, in agreement with other studies [39, 40].

Effect of Temperature

The effect of temperature on efficiency and benzene selectivity has been studied for different catalysts. Figure 7 shows how conversion varies with reaction temperature. The temperature curves exhibit linear character. For different catalysts, overall hexane conversion increased with temperature. The transformation of n-hexane on the tested catalysts started only above 250°C and the conversion reached 100% at 350°C - 400°C.

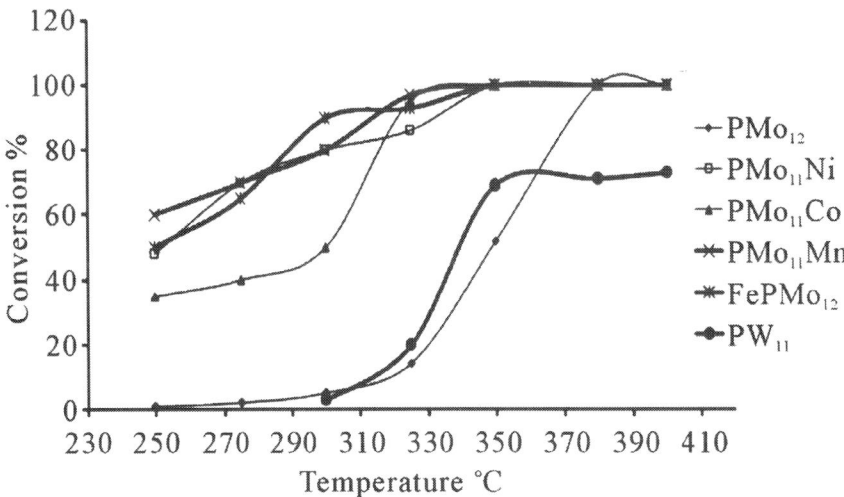

Figure 7: Conversion of n-hexane as function of temperature, (using N_2 as carrier gas).

Until 300°C, benzene was the unique or major product. The selectivity to benzene decreased with increasing temperature (Tables 4 and 5). The conversion increased from 2 up to 30%, whereas benzene selectivity decreased from 100% to 80% when the reaction temperature was increased from 300°C to 350°C when using $HPMo_{12}O_{40}$ heteropolyacid catalyst.

Methane was obtained with all catalysts at temperature above 300°C. Methane production selectivity increased with temperature at the expense of benzene. At 400°C n-hexane was converted completely to methane with all catalysts. The high methane selectivity was with PMoMn catalysts.

At 400°C the H_3PMo_{12} heteropolyacid was decomposed into Mo_3 and P_2O_5. This result agrees with literature [41] which showed that the isomerization of n-hexane was due to surface Bronsted acid sites.

Activation Energy

n-hexane conversion rate were determined over the temperature range 200°C - 450°C. From (Figures 8(a)-(f) n-hexane dehydrocyclization rate constants were determined over the temperature range 200°C - 450°C. From Arrhenius plots (Figure 9), Activation energies for the transformation of n-hexane over acidic catalysts were calculated as shown in Table 7.

Values of activation energies for the overall process of dehydrocyclization and aromatization of n-hexane using the acidic catalyst systems (43 to 95.2 kJ·mol⁻¹) are much lower than those known for metal type catalysts (typically 230 to 293 kJ·mol⁻¹) [42]. Acidic catalytic activity is expected to occur by the highly acidic hetero poly compounds. Actually the activation energies are comparable to those found for the transformation of the n-hexane in benzene using acidic catalysts. [43].

Furthermore, the selectivity of the benzene decreases, while the conversion increases, with higher temperatures or longer retention times. In case of all catalysts, the benzene is the secondary product of the reaction.

Table 7: Reaction activation energy for different catalyst systems measured based on amount of n-hexane consumption

Catalyst	Ea (KJ/mol)
$H_3PMo_{12}O_{40}$	95.2
$H(NH_4)_6PMo_{11}NiO_{40}$	83.2
$H(NH_4)_6PMo_{11}CoO_{40}$	63.8
$H(NH_4)_6PMo_{11}MnO_{40}$	43
$KFe_{0.1}PMo_{12}O_{40}$	83.2
$(TBA)_7PW_{11}O_{39}$	52.7

Effect of Time of the Reaction

The conversion of n-hexane was increase with time for all of the catalysts at 350°C as it is clear from Figure 10. The selectivity of benzene only slightly decreased with time on all catalysts Figure 8, when keeping temperature constant at 350°C This due to high thermodynamic stability of benzene.

Effect of Type of Carrier Gas

The type of carrier gas affected both conversion and selectivity of the reaction. The overall conversion increased by using H_2 as a carrier gas, while benzene selectivity increased when using N_2 as a carrier gas. This applied to all catalytic systems, as shown in Tables 5 and 6. Comparison between the two Tables shows that for each catalyst system, using H_2 gas gives higher conversion than using N_2. On the other hand, the Tables show using N_2 gave higher benzene selectivity than using H_2, for each catalyst.

Mechanism

A plausible mechanism has been suggested to explain the observations discussed above. As shown in Scheme I, two possible products can be expected by the HPA catalyst systems. The Scheme shows that with excess H_2 (carrier gas, as discussed above) the carbonium ion formed by super-acid activation of hexane will be cracked into CH_4 [44]. Such a process involves complete saturation of all carbon atoms with hydrogen, via complete cracking of the hydrocarbon chain. On the other hand, in case of nitrogen carrier gas (with no hydrogen) dehydrocylization process is favored to dominate, with no significant carbon chain cracking, as shown in the Scheme. The process is multi-stage involving β-H elimination in each step. This explains the production of cyclohexane, cyclohexene and cyclohexadiene as accompanying products to benzene.

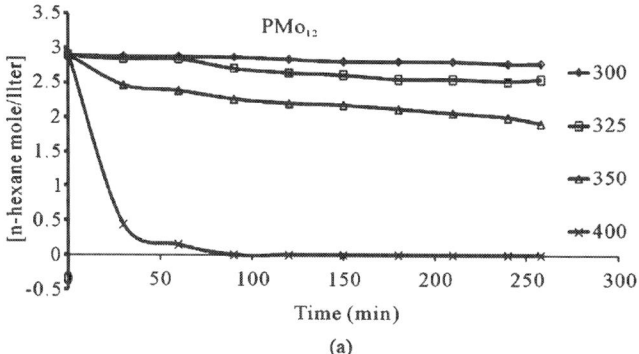

(a)

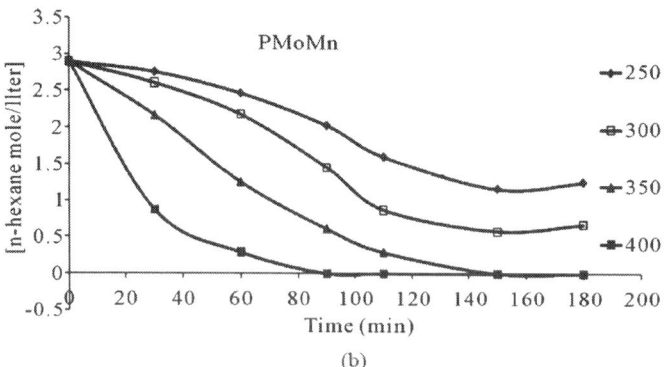

(b)

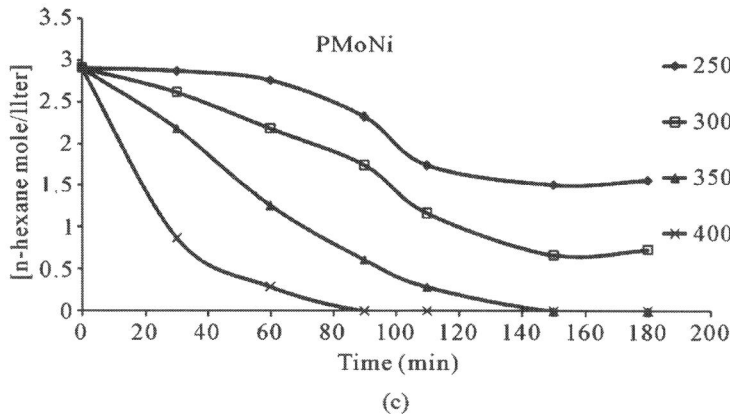

(c)

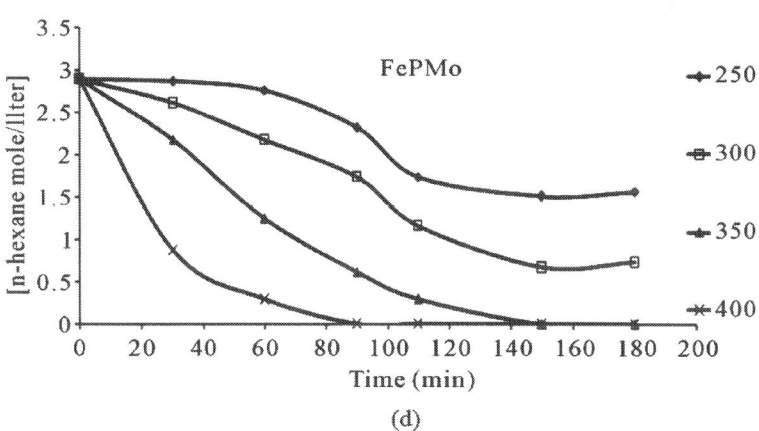

(d)

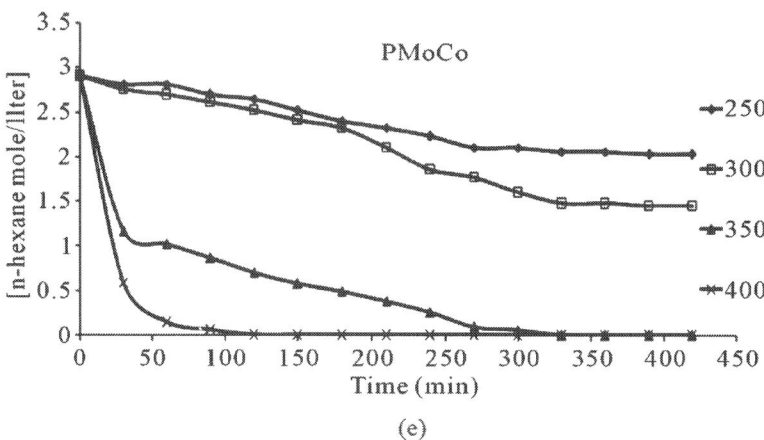

(e)

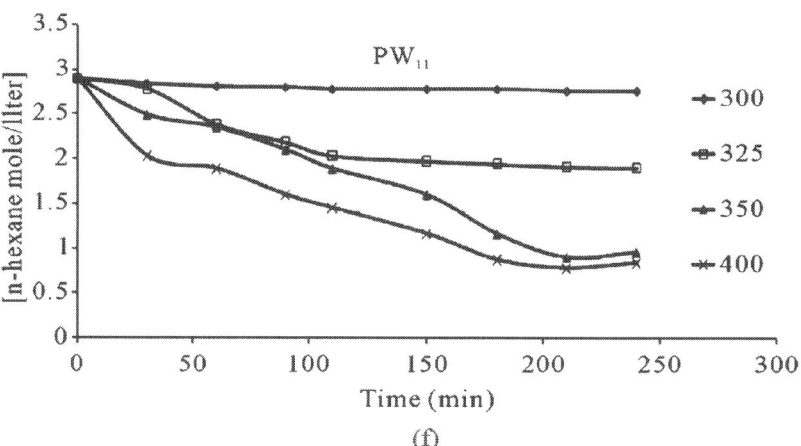

(f)

Figure 8: n-Hexane conversion as a function of time over: (a) PMo_{12} catalyst; (b) $PMo_{11}Mn$ catalyst; (c) $PMo_{11}Ni$ catalyst; (d) $FePMo_{12}$ catalyst; (e) $PMo_{11}Co$ catalyst; (f) PW_{11} catalyst.

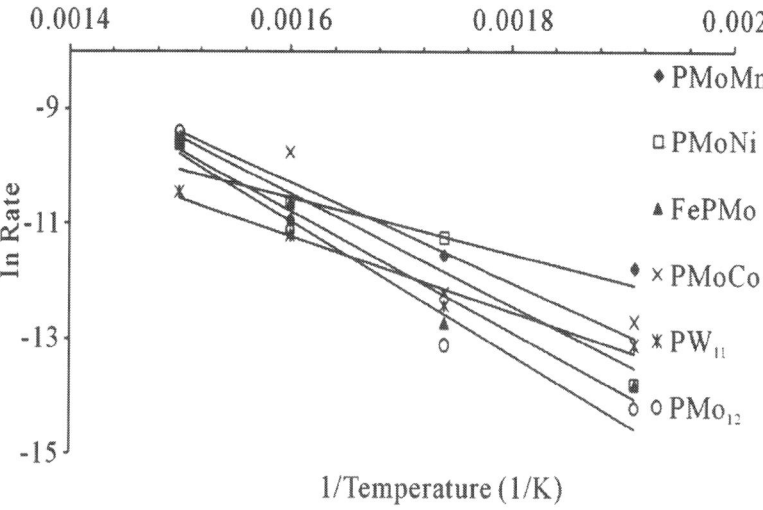

Figure 9: Arrhenuis plots for different catalyst systems (using N_2 as carrier gas).

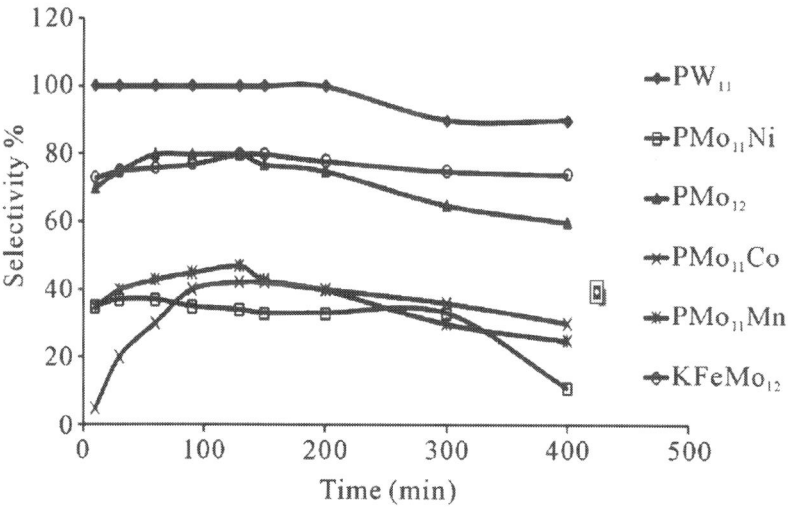

Figure 10: Selectivity of benzene as function of time at 350°C, (using N_2 as carrier gas).

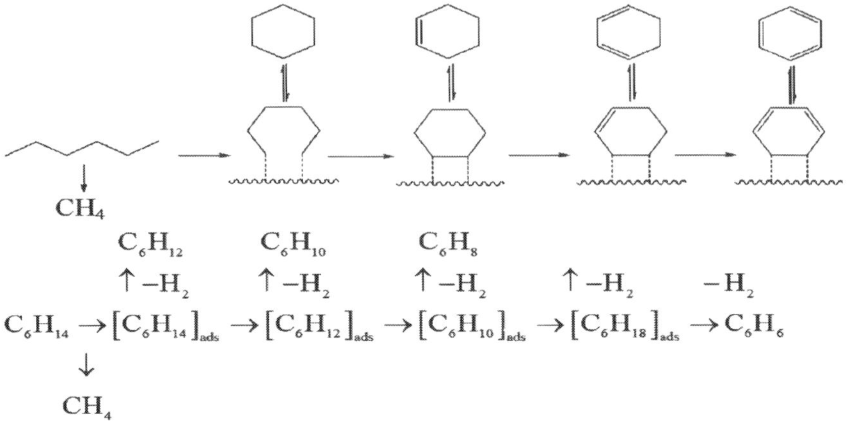

Figure 11: Reaction scheme of n-hexane dehydrocylization over heteropoly-compounds catalysts.

It should also be noted that this logic does not fully work under higher temperatures, where CH_4 production dominates even in the absence of hydrogen gas. In such a case, high temperature experiments without hydrogen yielded high molecular weight aromatics and tars, in parallel to CH_4 formation, which caused some technical difficulties such as reactor blockage during experiments. Therefore, if benzene is the desired product from nhexane, hydrogen gas should not be used, and the reaction must be conducted under mild temperatures (350°C or lower).

CONCLUSIONS

A number of molybdenum-based compounds $H_3PMo_{12}O_{40}$, $KFePMo_{12}O_{40}$, (TBA) $_7PW_{12}O_{40} \cdot xH_2O$, $(NH_4)_4PMo_{11}CoO_{40}$, $(NH_4)_4PMo_{11}NiO_{40}$, $(NH_4)_4PMo_{11}MnO_{40}$ have been prepared and characterized by UV, IR, NMR Spectra, TGA and XRD The compounds were of Keggin type structure. Due to their super-acidic nature, the compounds showed catalytic efficiency in the dehydrocyclization of n-hexane. The product distribution (benzene, cyclohexane, cyclohexene, cyclohexadiene and methane) strongly depends on the nature of the catalyst, the type of carrier gas and the reaction temperature. The observed differences in the behaviour of the studied catalysts should be firstly ascribed to the superacidic nature of the catalysts, which can activate saturated alkanes.

REFERENCES

1. B. C. Gates, "Catalytic Chemistry," John Wiley & Sons, Inc., New York, 1992.

2. A. Corma, J. M. Serra and A. Chica, "Discovery of New Paraffin Isomerization Catalysts Based on SO_4^{2-}/ZrO_2 and WO_x/ZrO_2 Applying Combinatorial Techniques," Catalysis Today, Vol. 81, No. 3, 2003, pp. 495-506. doi:10.1016/S0920-5861(03)00148-2

3. Y. Ono, "A Survey of the Mechanism in Catalytic Isomerization of Alkanes," Catalysis Today, Vol. 81, No. 1, 2003, pp. 3-16. doi:10.1016/S0920-5861(03)00097-X

4. B. Bachiller-Baeza, J. Alvarez-Rodríguez, A. GuerreroRuiz and I. Rodríguez-Ramos, "Support Effects on RuHPA Bifunctional Catalysts: Surface Characterization and Catalytic Performance," Applied Catalysis A: General, Vol. 333, No. 2, 2007, pp. 281-289.doi:10.1016/j.apcata.2007.09.027

5. M. H. Jordao, V. Simoes and D. Cardoso, "Zeolite Supported Pt-Ni Catalysts in n-Hexane Isornerization," Applied Catalysis A: General, Vol. 319, 2007, pp. 1-6.doi:10.1016/j.apcata.2006.09.039

6. Y. Gucbilmez, A. S. Yargic and I. Calis, "A Comparative Characterization of the HPA-MCM-48 Type Catalysts Produced by the Direct Hydrothermal and Room Temperature Synthesis Methods," Journal of Nanomaterials, Vol. 2012, 2012, Article ID: 210437.doi:10.1155/2012/210437

7. I. V. Kozhevnikov, "Heteropoly Acids and Related Compounds as Catalysts for Fine Chemical Synthesis," Catalysis Reviews, Vol. 37, No. 2, 1995, pp. 311-352.doi:10.1080/01614949508007097

8. I. V. Kozhevnikov, K. R. Kloetstra, A. Sinnema, H. W. Zandbergen and H. Van Bekkum, "Study of Catalysis Comprising Keteropoly Acid $H_3PW_{12}O_{40}$ Supported on MCM-41 Molecular Sieve and Amorphous Silica," Journal of Molecular Catalysis A: Chemical, Vol. 114, No. 1-3, 1996, pp. 287-298. doi:10.1016/S1381-1169(96)00328-7

9.   H. Hayashi and J. B. Moffat, "Methanol Conversion over Metal Salts of 12-Tungstophosphoric Acid," Journal of Catalysis, Vol. 81, No. 1, 1983, pp. 61-66.doi:10.1016/0021-9517(83)90146-X

10.   T. Baba, H. Watanabe and Y. Ono, "Generation of Acidic Sites in Metal Salts of Heteropoly Acids," Journal of Physical Chemistry, Vol. 87, No. 13, 1983, pp. 2406-2411.doi:10.1021/j100236a033

11.   K. Nowińska, "Catalytic Activity of Supported Heteropoly Acids for Reactions Requiring Strong Acid\Centres," Journal of Chemical Society, Faraday Transactions, Vol. 87, No. 5, 1991, pp. 749-753. doi:10.1039/ft9918700749

12.   V. Haensel, "Process of Reforming a Gasoline with an Alumina-Platinum-Halogen Catalyst," US Patent No. 24- 79109 and 2479110, 1949.

13.   H. E. Kluksdahl, "Reforming a Sulfur-Free Naphtha with a Platinum-Rhenium Catalyst," US Patent No. 3415737, 1968.

14.   M. H. Jordão, V. Simões and D. Cardoso, "Zeolite Supported Pt-Ni Catalysts in n-Hexane Isomerization," Applied Catalysis A: General, Vol. 319, 2007, pp. 1-6.doi:10.1016/j.apcata.2006.09.039

15.   C. Rocchioccioli-Deltcheff, M. Fournier, R. Franck and R. Thouvenot, "Vibration Investigatios of Polyoxometalates. 2. Evidence for Anion-Anion Interaction in Molybdenum (VI) and Tungsten(VI) Compounds Related to the Keggin Structure," Inorganic Chemistry, Vol. 22, No. 2, 1983, pp. 207-216. doi:10.1021/ic00144a006

16.   T. Mazari, S. Hocine, N. Salhi and C. Rabia, "Oxidation of Propane over Ammonium-Transition Metal Mixed Keggin Phosphomolybdate Salts," Journal of Natural Gas Chemistry, Vol. 19, No. 1, 2010, pp. 54-60.

17.   O. Benlounes, S. Cheknoun, S. Mansouri, C. Rabia and S. Hocine, "Catalytic Activation of C-H Bonds of Hydrocarbons by Heteropolycompounds," Journal of the Taiwan Institute of Chemical Engineers, Vol. 42, No. 1, 2011, pp. 132-137.

18.   F. M. Zhang, M. P. Guo, H. Q. Ge and J. Wang, "Hydroxylation of Benzene with Hydrogen Peroxide over Highly Efficient Molybdovanadophos Phoric Heteropoly Acid Catalysts," Chinese Journal of Chemical Engineering, Vol. 15, No. 6, 2007, pp. 895-

898.doi:10.1016/S1004-9541(08)60021-X

19. O. Benlounesa, S. Mansouria, C. Rabiab and S. Hocine, "Direct Oxidation of Methane to Oxygenates over Heteropolyanions," Journal of Natural Gas Chemistry, Vol. 17, No. 3, 2008, pp. 309-312. doi:10.1016/S1003-9953(08)60070-5

20. H. Kima, P. Kima, K. Leeb, S. H. Yeomc, J. Yia and I. K. Song, "Preparation and Characterization of Heteropoly Acid/ Mesoporous Carbon Catalyst for the Vapor-Phase 2- Propanol Conversion Reaction," Catalysis Today, Vol. 111, No. 3-4, 2006, pp. 361-365.doi:10.1016/j.cattod.2005.10.048

21. J. N. Beltramini, "Studies in Surface Science and Catalysis," Proceedings of the 3rd International Mesostructured Materials Symposium, Nanotechnology in Mesostructured Materials, Jeju, 8-11 July 2002, pp. 653-656.

22. R. S. Drago, J. A. Dias and T. O. Maier, "An Acidity Scale for Brönsted Acids Including $H_3PW_{12}O_{40}$," Journal of American Chemical Society, Vol. 119, No. 33, 1997, pp. 7702-7710. doi:10.1021/ja9639123

23. K. Nowinska, R. Fiedorow and J. Adamiec, "Catalytic Activity of Supported Heteropoly Acids for Reactions Requiring Strong Acid Centres," Journal of the Chemical Society, Faraday Transactions, Vol. 87, No. 5, 1991, pp. 749-753. doi:10.1039/ft9918700749

24. K. J. Nowinska, "Evidence for Superacid Sites on the Ammonium Salt of 12-Tungstophosphoric Acid from a Catalytic Test Reaction," Journal of the Chemical Society, Chemical Communications, 1990, pp. 44-45. doi:10.1039/c39900000044

25. C. Marchal-Roch, J. M. Millet and C. R. Acad, "Phosphomolybdic Heteropolycompounds as Oxidation Catalysts. Effect of Transition Metals as Counter-Ions," Science Chemistry, Vol. 4, No. 5, 2001, pp. 321-329.

26. S. Hocine, C. Rabia, M. M. Bettahar and M. Fournier, "Oxidative Dehydrogenation of Cyclohexane over Heteropolymolybdates," Studies in Surface Science and Catalysis, Vol. 130, 2000, pp. 1895-1900. doi:10.1016/S0167-2991(00)80478-4

27. R. Tayebee, "Simple Heteropoly Acids as Water-Tolerant Catalysts in the Oxidation of Alcohols with 34% Hydrogen Peroxide," Journal of the Korean Chemical Society, Vol. 52, No. 1, 2008,

pp. 23-29. doi:10.5012/jkcs.2008.52.1.023

28. T. Okuhara, "Catalytic Chemistry of Heteropoly Compounds," Advances in Catalysis, Vol. 41, 1996, pp. 113- 252. doi:10.1016/ S0360-0564(08)60041-3

29. Y. He, W. Sang, J. Wang, R. Wu and J. Min, "Vertically Well-Aligned ZnO Nanowires Generated with Self-Assembling Polymers," Materials Chemistry and Physics, Vol. 94, No. 1, 2005, pp. 29-33.

30. C. L. Hill and C. M. Prosser-Mccartha, "Homogeneous Catalysis by Transition Metal Oxygen Anion Clusters," Coordination Chemistry Reviews, Vol. 143 ,1995, pp. 407- 455.

31. A. V. Churakov, E. A. Legurova, A. A. Dutov, P. V. Prikhodchenko and T. A. Tripol'skaya, "Peroxide Derivatives of Heteropoly Compounds with Keggin Anions $[PW_{12}O_{40}]^{3-}$ and $[SiW_{12}O_{40}]^{4-}$: Synthesis and Structure," Russian Journal of Inorganic Chemistry, Vol. 53, No. 8, 2008, pp. 1187-1192. doi:10.1134/ S0036023608080068

32. T. J. R. Weakley and S. A. Malik, "Triheteropolyanions Containing Copper(II), Manganese(II), or Manganese (III)," Journal of Inorganic and Nuclear Chemistry, Vol. 32, No. 12, 1970, pp. 3875-3890.

33. O. Benlounesa, "Oxidation of Methane on Heteropoly Compounds," Ph.D. Dissertation, University of Mouloud Mammeri-Tizi-Ouzou, Tizi-Ouzou, 2010.

34. S. S. Lima, G. I. Park, I. K. Song and W. Y. Lee, "Heteropolyacid (HPA)-Polymer Composite Films as Catalytic Materials for Heterogeneous Reactions," Journal of Molecular Catalysis A: Chemical, Vol. 182-183, 2002, pp. 175-183. doi:10.1016/ S1381-1169(01)00464-2

35. S. Hocine, "Heteropolyphosphomolybdates. Preparation, Charactarization, Catalytic Activity of Cyclohexane Oxydehydrogenation," Ph.D. Dissertation, Houari Boume Diene University, Algiers, 2003.

36. A. Popa, V. Sasca, M. Stefanescu, E. Kis and R. Marinkovic-Neducin, "The Influence of the Nature and Textural Properties of Different Supports on the Thermal Behavior of Keggin Type Heteropolyacids," Journal of the Serbian Chemical Society, Vol. 71, 2006, pp. 235-249. doi:10.2298/JSC0603235P

37. G. A. Tsigdinos and C. J. Hallada, "Molybdovanadophosphoric Acids and Their Salts Investigation of Methods of Preparation and Characterization," Inorganic Chemistry, Vol. 7, No. 3, 1968, pp. 437-441. doi:10.1021/ic50061a009

38. T. Okuhara, N. Mizuno and M. Misono, "Catalysis by HeteroPolycompounds—Recent Developments," Applied Catalysis A: General, Vol. 222, No. 1-2, 2001, pp. 63-77. doi:10.1016/S0926-860X(01)00830-4

39. R. Hubaut, B. Ouled Ben Tayeb, W. Kuang, A. Rives and M. Fournier, "Mechanical Mixtures of Me (Ni, Pd)Ce Oxides and Silica-Supported Heteropolyacids: Role and Optimal Content of Each Active Species for n-Hexane Isomerization," Kinetics and Catalysis, Vol. 47, No. 1, 2006, pp. 20-24. doi:10.1134/S0023158406010046

40. V. V. Brei, O. V. Melezhyk, S. V. Prudius, M. M. Levechuk and K. I. Patrylak, "Superacid WO_x/ZrO_2 Catalysts for Isomerization of n-Hexane and for Nitration of Benzene," Studies in Surface Science and Catalysis, Vol. 143, 2002, pp. 387-395. doi:10.1016/S0167-2991(00)80679-5

41. B. Demirel and E. N. Givens, "Transformation of Phosphor Molybdic Acid into an Active Catalyst with Potential Application in Coal Liquefaction," Catalysis Today, Vol. 50, No. 1, 1999, pp. 149-158. doi:10.1016/S0920-5861(98)00472-6

42. F. Garin and F. G. Gault, "Mechanisms of Hydrogenolysis and Isomerization of Hydrocarbons on Metals. VIII. Isomerization of Carbon-13 Labeled Pentanes on a 10% Platinum-Aluminum Oxide Catalyst," Journal of the American Chemical Society, Vol. 97, No. 16, 1975, pp. 4466- 4476. doi:10.1021/ja00849a004

43. F. R. Ribeiro, C. Marcilly and M. Guisnet, "Hydroisomerization of n-Hexane on Platinum Zeolites," Journal of Catalysis, Vol. 78, No. 2, 1982, pp. 267-280. doi:10.1016/0021-9517(82)90311-6

44. S. Kotrel, H. Knozinger and B. C. Gates, "The HaagDessau Mechanism of Protolytic Cracking of AlKanes," Microporous and Mesoporous Materials, Vol. 35-36, 2000, pp. 11-20. doi:10.1016/S1387-1811(99)00204-8

Tool Support for the Management of Design Processes in Chemical Engineering

Manfred Nagl[a], Bernhard Westfechtel[a], and Ralph Schneider[b]

[a]RWTII Aachen, Lehrstuhl für Informatik III, D-52056 Aachen, Germany

[b]RWTH Aachen, Lehrstuhl für Prozesstechnik, D-52056 Aachen, Germany

ABSTRACT

Design processes in chemical engineering are hard to support. In particular, this applies to conceptual design and basic engineering, in which the fundamental decisions concerning the plant design are performed. The design process is highly creative, many design alternatives are explored, and both unexpected and planned feedback occurs frequently. As a consequence, it is inherently difficult to manage design processes, i.e. to coordinate the effort of experts working on tasks such as creation of flow diagrams, steady-state and dynamic simulations, etc. On the other hand, proper management is crucial because of the large economic impact of the performed design decisions. We present a management system which takes the difficulties mentioned above into account by supporting the coordination of dynamic design processes. The management system equally covers products, activities, and resources, and their mutual relationships. With respect to coverage and integration, and with respect to the dynamics of design processes, the functionality of the management system goes considerably beyond commercial project, document, and workflow management systems.

INTRODUCTION AND MOTIVATION

Design processes in engineering disciplines, like chemical engineering, often deliver good results, but sometimes perform less effective than they could. Reasons, among others, are that neither the process is explicitly and clearly structured nor its complex result. Especially, the experience of designers is not explicitly gathered and, therefore, cannot be used, the many mutual relationships between parts of the design product are neither explicitly stored nor maintained, designers may interpret the design in different ways, and the management of a project is not given a clear view about the state of the project at a certain time.

Correspondingly, there is a *lack* of *semantic support* by current *tools* for collaborative design processes, carried out by different persons, with different roles, on different sites, eventually in different companies, which altogether build up or maintain the complex product of a design process. Moreover, there are lot of *gaps* with respect to tool support in a collaborative design process. As a consequence, the vision of an integrated design environment providing high-level tools has been realized only to a limited extent.

The *Collaborative Research Center476IMPROVE* (Nagl and Westfechtel, 1998 and Marquardt and Nagl, 1999), is an integrated project staffed by chemical engineers, plastic engineers, ergonomics researchers, and different groups from computer science (software engineering, information systems, communication systems) with the aim of solving the problems described above. The project concentrates on the early phases of process engineering, namely conceptual process design and basic engineering, the tasks of which are to structure, to simulate, and to evaluate a plant design under different perspectives. This part of the overall design process is especially challenging from a research perspective (many creative decisions, permanent changes, study of variants etc.).

With respect to the development of an integrated design environment, IMPROVE follows a mixed *top–down/bottom–up* approach (Fig. 1). Bottom–up means that existing tools and platforms are re-used as far as possible (grey regions). For example, we make use of existing tools for performing steady-state or dynamic simulations. To further improve the functionality offered to the designers, new tools and services are added. The tools are designed in such a way that they fit into the overall architecture (top–down approach).

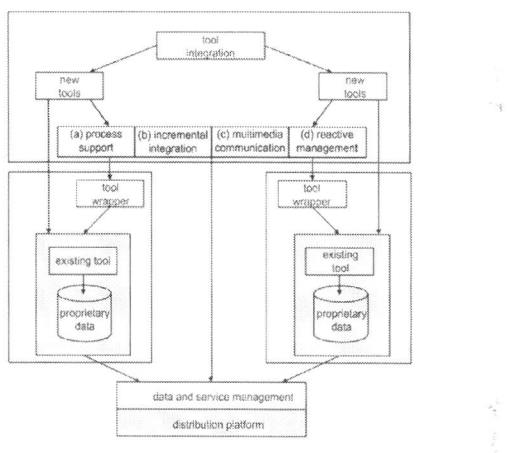

Figure 1: Mixed tool integration approach (coarse architectural sketch).

Within the IMPROVE project, there are four subprojects which address improved tool support in different functional areas. Within these subprojects, tool services are developed which may be used for the implementation of design tools with innovative functionality (Fig. 1(a)–(d)):

- Fine-grained *process support tools* (Pohl et al., 1999) aim at supporting designers in their interaction with design tools by process fragments which (partly) automate sequences of command invocations.

- *Incremental integration tools* (Becker, Haase, Westfechtel, & Wilhelms, 2002) assist designers in keeping inter-dependent design data consistent with each other (e.g., flow diagrams and simulation models).

- Using *multi-media communication tools* (Schüppen, Trossen, & Wallbaum, 2002), designers may discuss and resolve design problems in joint working sessions (conferences).

- *Reactive management tools* Westfechtel, 1999 assist in managing design processes at a more coarse-grained level than the process support tools mentioned above.

These *new functionalities* work *synergetically together*. To take one example, let us regard the change of a flow diagram, for which process support (a) can be used. The dependent tasks are determined by reactive management (d). For changing the dependent documents (e.g., a simulator input description), integration tools (b) are applied. During the change subprocess spontaneous multimedia communication is used as well as a multimedia conference (c) at the end of the subprocess.

In this paper we *concentrate* on tool support for the *management* of *design processes* in chemical engineering (i.e. (d) in Fig. 1). Management, thereby, is restricted to the *coordination* of a design project. So, no strategic or psychological aspects are discussed. However, all aspects of coordination (products, activities, and resources) are seen as being tightly integrated. Since the design process changes permanently, coordination has to react accordingly (dynamics). Management is supported by a collection of tools which constitute a reactive management system. This system is called *A*daptable and *H*uman-Centered *E*nvironment for the M*A*nagement of *D*evelopment Processes (*AHEAD* Jäger, Schleicher, and Westfechtel, 1999a and Westfechtel, 1999).

Management in the sense of this paper is closely *related* to *major design decisions* of a chief designer. So, e.g., choosing one of different variants for a part of the plant induces a specific form of the design result (how many and which documents), design process (which design tasks) and corresponding required resources. Conversely, a limit in the design process' costs influences how intensively design variants can be studied. So, both levels, major design decisions on the technical level and coordination of the design process on the managerial level, are mutually dependent.

The rest of this paper is structured as follows: In Section 2, we elaborate on the features of design processes in chemical engineering. We also introduce a case study, namely the design of a plant for polyamide6. In Section 3, we discuss the management of design processes and the current state of the art of tool support. We demonstrate that commercial tools for workflow, project, or document management suffer from several shortcomings, e.g., limited support for dynamically evolving design processes. InSection 4, we present our management system (AHEAD) and explain in what respects it goes beyond the functionality of current commercial systems. In Section 5, we demonstrate its use by applying it to the case study introduced in Section 2. While the management system incorporates novel functionality, it is a research prototype which cannot be immediately applied to industrial practice. Therefore, we indicate different ways of technology transfer in Section 6. Finally, we summarize our contributions and describe current and future work in Section 7.

DESIGN PROCESSES IN CHEMICAL ENGINEERING

From an economic point of view the early phases of design processes, namely conceptual process design and basic engineering, are worth considering. The decisions made in these phases have a great impact on the later ones. McGuire and Jones (1989) report that up to 80% of the capital costs of a plant are determined in conceptual process design. The importance of design processes in (chemical) engineering and the need for improving and modeling them has been stated in the literature by several authors (e.g.,Mostow, 1985, Bañares-Alcántara, 1995, Ponton, 1995 and Westerberg, Subrahmanian, Reich, Konda, and n-dim group, 1997).

Characteristics of Design Processes

The life cycle of a chemical design process ranges from the definition of a design objective to the construction of the plant as shown in Fig. 2 (developed within the Global CAPE-OPEN projectBraunschweig & Gani, 2002). Besides the performed *activities* (e.g., basic engineering), which are assigned to roles or departments (*resources*), the needed and produced information (*products*) characterize the design process and are represented in the figure. The temporal order of the activities is given by the control flow (solid lines), the flow of information by the dashed lines.

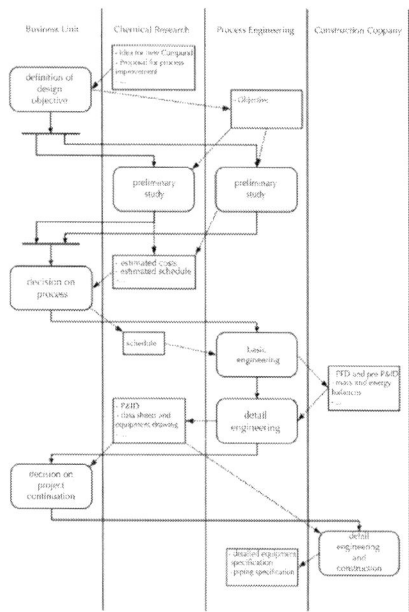

Figure 2: Life cycle of a chemical design process.

Our management system aims at supporting the engineering design process (namely the activities from the start of the project to the ones including the detailed process design) with special consideration for the characteristics of a (chemical) engineering design process:

- *Designers:* The designers have different backgrounds and work together in a relatively small, but interdisciplinary team (up to 10 people), lead by one technical project manager.
- *Location:* The design team can be located in different departments and enterprises, distributed across different sites all over the world.
- *Duration:* Chemical engineering design projects differ in their duration, depending on the goal of the design project (e.g., design of a complete new plant, or just a retrofit). They last from several months up to many years.

- *Creativity:* Design processes include highly creative work processes. They are ill-defined and very complex, therefore very difficult to plan in advance.

- *Iterations:* There are a lot of iterations in a design process. Many activities have to be performed several times at different levels of detail (granularity) depending on the available information.

- *Feedback:* The activities performed during a design process are highly interconnected. Therefore changes made in a later phase of the design process can lead to unwanted feedback to activities already performed earlier during process design.

- *Documentation:* The exchange of information between the designers is mainly done during project meetings and by the exchange of email and paper. There is almost no reuse of successful problem solutions of previous design projects due to the difficulties in exchanging and using consistent information as well as missing or bad documentation.

- *Alternatives*: During a design process many different alternatives are created. The detailed investigation of each alternative may lead to totally different design processes.

- *Software tools*: A multiplicity of software tools is used by the designers. These tools often have incompatible data formats which make the exchange of information very difficult or even impossible. Large amounts of data and documents are produced by the software tools which have to be managed.

- *Design product:* The product of a design process in chemical engineering is the plant design. Chemical plants are unique and no bulk products.

In particular, these characteristics imply that design processes cannot be planned in detail in advance because they are dynamically changing due to the highly creative character of the processes and the interdependencies between the different activities and products. Effective support for managing design processes can only be enabled by the combined consideration of all these aspects.

Polyamide6 Case Study

The case study given here serves different purposes. First of all, we want to understand the workflow of industrial design processes in order to identify weak points and define requirements for the development of new tool functionalities respectively new tools. This case study can therefore be seen as a guideline for our tool design process. It is a common basis for our work in the IMPROVE project. All tools developed in IMPROVE are tested in the context of the case study. Furthermore, these tools are integrated in a common prototype demonstrating the interactions between the different tools and their support functionalities. Following this working procedure it is possible to evaluate whether the tools really fulfill the defined requirements and contribute significantly to an improvement of design processes in chemical engineering.

In order to demonstrate and evaluate the functionalities of the management system, a realistic *case study* is needed. Since such a case study is not available in the literature, we developed one by ourselves.

First of all we tried to record and structure design processes in chemical engineering at a very coarse level (independently from a concrete example) based on literature (Rudd, Powers, and Siirola, 1973, Douglas, 1988, Biegler, Grossmann, and Westerberg, 1997 and Blass, 1997), the PIEBASE activity modelPIEBASE Working Group 2, (1998), and own experiences. Furthermore, we developed a process for the production of *polyamide6* (nylon6) fulfilling a specified design task. During this development we observed ourselves and recorded our activities in the form of activity diagrams. In parallel to this we conducted five interviews with project managers of an industrial partner. In these interviews the interviewee had to remember in the sense of a case study approach (Yin, 1984) a past design process. These three elements (literature, self observation, and interviews) form the basis on which our case study is built.

The case study describes the design process of a chemical plant for the production of polyamide6 including the polymer compounding and post-processing. It focuses on the workflow, the people involved and the tools used together with their interactions. The task given in this case study is to design a plant for producing 40 000 tons polyamide6 per year with a given product quality. Polyamide6 is produced by the polymerization of -caprolactam. There are two possible reaction mechanisms, the hydrolytic and the anionic polymerization (Kohan, 1995). Since the anionic polymerization is mainly used for special polymers, this case study focuses on the *hydrolytic polymerization*, which is also more often applied industrially. This polymerization consists of three single reaction steps: ring opening of -caprolactam, poly-condensation, and poly-addition. The case study covers the design of the reaction and separation system as well as the extrusion. A block flow diagram representing the continuous polymerization of polyamide6 is shown in Fig. 3.

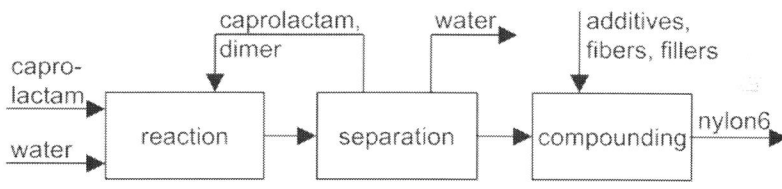

Figure 3: Block flow diagram of polyamides6 process.

There are different kinds of *reactors* that can be used for the polymerization: sequences of two or more tank reactors or plug flow reactors (Gerrens, 1981) and a special reactor developed for the polyamide6 production, the VK-tube (Deutsches Patentamt, 1969). The polymer melt at the outlet of the reactor contains monomer, oligomers, and water, which must be removed in order to meet the required product quality. Therefore, a *separation* is needed. Two separation mechanisms can be used here: evaporation in a wiped-film-evaporator or extraction with water to remove -caprolactam with a successive drying step (Kohan, 1995). Polymer post-processing is done within an extruder: additives, fillers, and fibers are added in order to meet the specified product qualities. Within the extruder, an additional polymerization of the melt and the degassing of volatile components are possible.

For all process units, mathematical models are developed and used for simulation within different simulators. Because the design of each distinct process unit requires very specific knowledge, each block is studied in detail by different experts in our case study.

The different alternatives for reaction, separation, and extrusion lead to several alternative processes for the polyamide6 production. In Fig. 4, a flow diagram of one process alternative is given: the reaction takes place within two reactors, separation is done by leaching and drying of polymer pellets. The cleaned pellets are re-melted in the extruder so that additives can be added. More detailed descriptions of the case study can be found in Bayer, Eggersmann, Gani, and Schneider (2002) and Eggersmann, Hackenberg, Marquardt, and Cameron (2002).

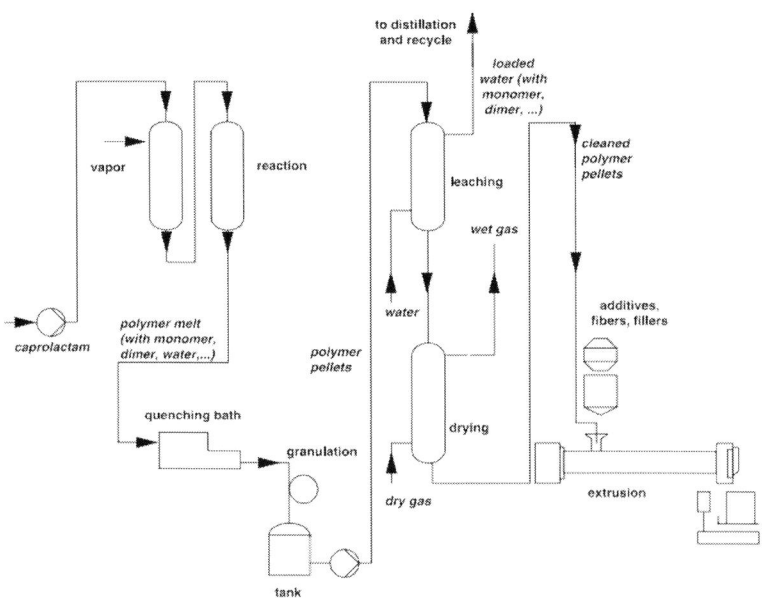

Figure 4: Flow diagram for polyamide production.

In Fig. 5, a simplified overview on the polyamide6 case study is given from a workflow perspective. Regarding the above mentioned applications of our case study it was important to choose a suitable notation, in the sense that engineers as well as computer scientists are able to understand the modeled content and that all necessary information is included in such a workflow model. The notation used is the*C3 formalism* (Foltz, Killich, Wolf, Schmidt, & Luczak, 2001), a modeling language for the notation of work processes, based on the Unified Modeling Language (uml) (Booch, Rumbaugh, & Jacobson, 1999). The abbreviation C3 stands for the three aspects of workflow modeling which are represented in this formalism: cooperation, coordination, and communication. The elements of C3 are roles (e.g., simulation expert), activities (e.g., design reaction alternatives), input/ output information (not shown in this figure), control flows (solid lines including forks and joins represented by bars), information flows (also not shown), and synchronous communication (represented by a filled square).

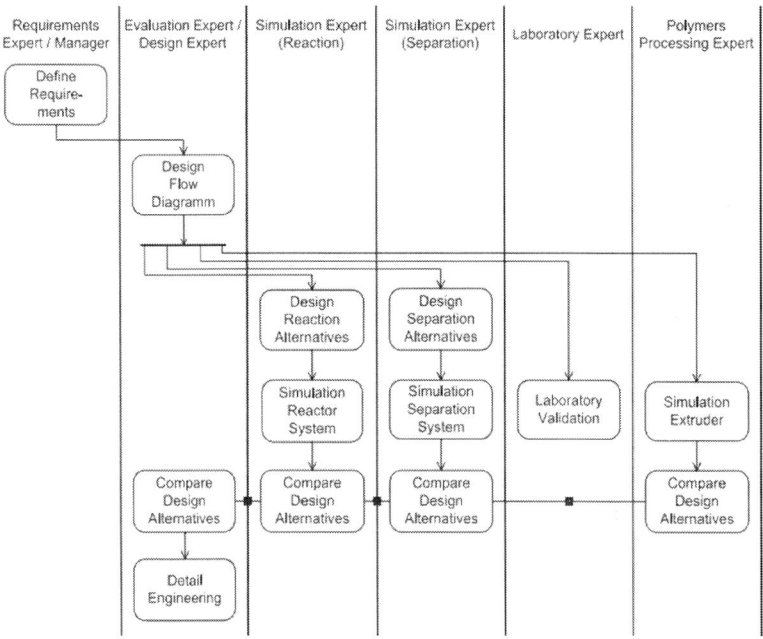

Figure 5: Simplified overview on case study.

After the start of the project, different alternatives are evaluated on the basis of block flow diagrams. Reaction, separation, and extrusion are investigated in parallel within different simulation tools. The simulation results are compared with experimental results leading to an improvement and refinement of the used mathematical models. After the completion of these activities a plant concept is determined. The case study in its actual form represents one part of an industrial design process, namely the early phases of basic engineering.

On the basis of this case study (with about one hundred activities) the management system has been developed and tested. For reasons of clarity only a small part of the case study is presented and used for demonstration in the following sections.

MANAGEMENT OF THE DESIGN PROCESS

Basic Notions

In the previous section, we have introduced our domain of discourse—design processes in chemical engineering—and a case study (design of a plant for producing polyamide6) which serves as a reference scenario within the IMPROVE project. In this paper, we are specifically concerned with the management of design processes. In Section 2.1, we have described several features of design processes which challenge the capabilities of management. In particular, design processes are highly creative, many iterations are necessary, the tasks to be solved are not known beforehand, design processes last for a long time span, experts from different disciplines have to cooperate smoothly, these experts use heterogeneous tools, and they have to document the results of their work in a traceable way.

Before we discuss solutions to these problems, we have to introduce a set of basic notions which we will use throughout the rest of this paper. In general terms, *management* can be defined as 'all the activities and tasks undertaken by one or more persons for the purpose of planning and controlling the activities of others in order to achieve an objective or complete an activity that could not be achieved by the others acting alone' (Thayer, 1988). This definition stresses coordination as the essential function of management.

More specifically, we focus on the management of design processes by coordinating the technical work of designers. We do not target senior managers who work at a strategic level and are not concerned with the details of enterprise operation. Rather, we intend to support project managers who collaborate closely with the designers performing the technical work. Such managers, who are deeply involved in the operational business, need to have not only managerial but also technical skills ('chief designers').

The distinction between *persons* and *roles* is essential: when referring to a 'manager' or a 'designer', we are denoting a role, i.e. a collection of authorities and responsibilities. However, there need not be a 1:1 mapping between roles and persons playing roles. In particular, each person may play multiple roles. For example, in chemical engineering it is quite common that the same person acts both as a manager coordinating the project and as a (chief) designer who is concerned with technical engineering tasks.

In order to support managers in their coordination tasks, design processes have to be dealt with at an appropriate level of detail. We may roughly distinguish between three levels of *granularity*:

- At a *coarse-grained level*, design processes are divided into phases (or working areas) according to some life cycle model (Fig. 2).

- At a *medium-grained level*, design processes are decomposed further down to the level of documents or tasks, i.e. units of work distribution.

- At a *fine-grained level*, the specific details of design subprocesses are considered. For example, a simulation expert may build up a simulation model from mathematical equations.

Given our understanding of management as explained above, the coarse-grained level does not suffice; rather, decomposition has to be extended to the medium-grained level. On the other hand, management is usually not interested in the technical details of how documents are structured or how the corresponding personal subprocess is performed. Thus, the *managerial level*, which defines how management views design processes, comprises both coarse- and medium-grained representations.

In order to support managers in their coordination tasks, they must be supplied with appropriate views (abstractions) of design processes. Such views must be comprehensive inasmuch as they include products, activities, and resources (and their mutual relationships):

The term *product* denotes the results of design subprocesses (e.g., flow diagrams, simulation models, simulation results, cost estimates, etc.) [1]. These may be organized into *documents*, i.e. logical units which are also used for work distribution or version control.

- The term *activity* denotes an action performing a certain function in a design process. At the managerial level, we are concerned with *tasks*, i.e. descriptions of activities assigned to designers by managers.

- Finally, the term *resource* denotes any asset needed by an activity to be performed. This comprises both human and computer resources (i.e. the designers and managers participating in the design process as well as the computers and the tools they are using).

Thus, an overall *management configuration* consists of multiple parts representing products, activities, and resources. An example is given in Fig. 6. Here, we refer to the polyamide6 design process introduced earlier. On the left, the figure displays the roles in the design team as well as the designers filling these roles [2]. The top region on the right shows design activities connected by control and data flows. Finally, the (versioned) products of these activities are located in the bottom–right region.

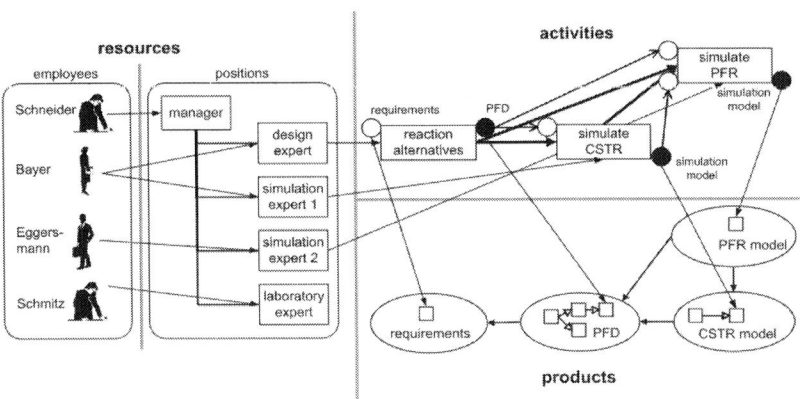

Figure 6: Management configuration.

Below, we give a more detailed description of Fig. 6:

- *Products.* The results of design processes such as process flow diagrams (PFDs), steady-state and dynamic simulations, etc. are represented by documents (ellipses). Documents are interdependent, e.g., a simulation model depends on the PFD to which it refers (arrows between ellipses). The evolution of documents is captured by version control (each box within an ellipsis represents a version of some document).

- *Activities.* The overall design process is decomposed into tasks (rectangular boxes) which have inputs and outputs (white and black circles, respectively). The order of tasks is defined by control flows (thick arrows); e.g., reaction alternatives must have been inserted into the flow diagram before they can be simulated. Finally, data flows (arrows connecting circles) are used to transmit document versions from one task to the next.

- *Resources.* Employees (icons on the left) such as Schneider, Bayer, etc. are organized into project teams which are represented by organization charts. Each box represents a position, lines reflect the organizational hierarchy. Employees are assigned to positions (or roles). Within a project, an employee may play multiple roles. e.g., Mrs Bayer acts both as a designer and as a simulation expert in the polyamide6 team.

- *Integration.* There are several relationships between products, activities, and resources. In particular, tasks are assigned to positions (and thus indirectly to employees). Furthermore, document versions are created as outputs and used as inputs of tasks.

It is crucial to understand the scope of the term 'management' as it is used in this paper. As already stated briefly above, management requires a certain amount of abstraction. This means that the details of the*technical level* are not represented at the managerial level. This is illustrated in Fig. 7, whose upper part shows a small cutout of the management configuration of Fig. 6. On the managerial level, the design process is decomposed into activities such as creation of reaction alternatives and simulation of these alternatives. Activities generate results which are stored in document versions. At the managerial level, these versions are basically considered black boxes, i.e. they are represented by a set of descriptive attributes (author, creation date, etc.) and by references to the actual contents, e.g., PFDs and simulation models. How a PFD or a simulation model is structured internally (and how their contents are related to each other), goes beyond the scope of the managerial level. Likewise, the managerial level is not concerned with the detail personal process which is executed by some human to create a PFD, a simulation model, etc.

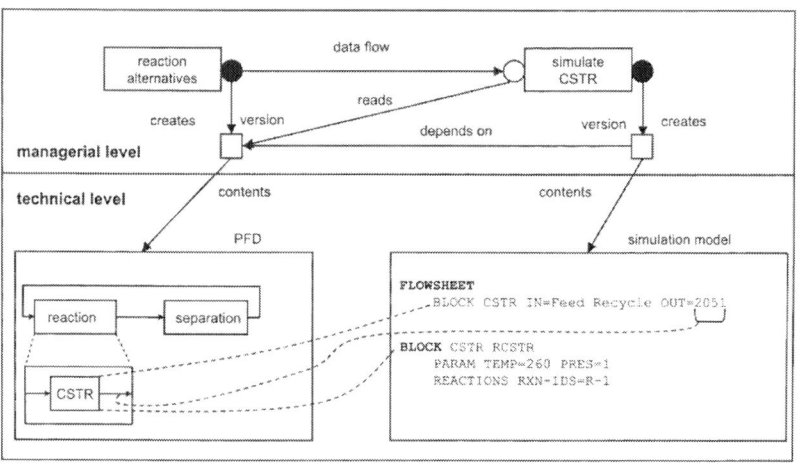

Figure 7: Managerial and technical level.

This does not imply that technical details are ignored. Rather, it must be ensured that the managerial level actually constitutes a correct abstraction of the fine-grained information at the technical level—and also controls technical activities. In fact, the management system described in this paper is part of an integrated environment for supporting design processes in chemical engineering (see also Fig. 1). As such, it is integrated with tools providing fine-grained product and process support (Bayer et al., 2002; Becker et al., 2002). The interplay of the tools of the overall environment is sketched only briefly in this paper; see (Nagl, Schneider, & Westfechtel, in press).

Particular attention has to be paid to the *dynamics* of design processes (called evolution in (Smithers & Troxell, 1990)). As we have demonstrated in the previous section, the design process is not known in advance. Rather, it continuously evolves during execution. Often, the term 'dynamics' is interpreted in a rather restricted way, referring only to the execution of a known process with a static definition. In addition, we have to take care of evolving definitions, i.e. the activities to be executed as well as their relationships are defined only at runtime.

As a consequence, all parts of a management configuration evolve continuously:

- *Products.* The product structure is determined only during the design process. It depends on the flow diagram which is continuously extended and modified. Other documents such as simulation models and simulation results depend on the flow diagram. Moreover, different variants of the chemical process are elaborated, and selections among them are performed according to feedback gained by simulation and experiments.

- *Activities.* The activities to be performed depend on the product structure, feedback may require the re-execution of terminated activities, concurrent/simultaneous engineering calls for sophisticated coordination of related activities, etc.

- *Resources.* Resource evolution occurs likewise: new tools arrive, old tool versions are replaced with new ones, the project team may shrink due to budget constraints, or it may be extended to meet a crucial deadline, etc.

However, a management configuration should not evolve in arbitrary ways. There are *domain-specific constraints* which have to be met. In particular, activities can be classified into types such as requirements definition, design, simulation, etc. (likewise for products and resources). Furthermore, the way how activities are connected is constrained as well. For example, a flow diagram can be designed only after the requirements have been defined. Such domain-specific constraints should be taken into account such that they restrict the freedom of evolution.

Management Systems: State of the Art

Above, we have discussed design processes and their management on a conceptual level. In the following, we will be concerned with *tool support* for managing design processes. From the previous discussion, we derive a set of crucial requirements for management tools for design processes [3]:

- *Medium-grained representation.* The management of design processes has to be supported at an appropriate level of detail. As we have argued before, this requires a medium-grained representation of the design process.

- *Coverage and integration at the managerial level.* Management tools have to deal equally with products, activities, and resources. In addition, the relationships among them have to be taken into account.

- *Integration between managerial and technical level.* Managerial activities have to be coupled with technical activities. Practically speaking, this implies e.g., that designers have to be supplied with the documents they are going to manipulate, as well as with the tools they are going to use.

- *Dynamics of design processes.* Design processes evolve continuously during execution. Thus, a management system must support dynamic changes so that product evolution, feedback, simultaneous and concurrent engineering etc. can be expressed.

- *Adaptability.* Management tools have to be adapted to a specific application domain. For example, they must be aware of the types of activities performed in design processes in chemical engineering, and they must provide domain-specific operations to their users.

In industry, a large variety of commercial systems are being used for the management of design processes. These include systems for project management, workflow management, and product management, which are discussed in turn below. All of these systems meet the requirements stated above only partially (Table 1[4]).

Table 1: Comparison of AHEAD with commercial management systems

	AHEAD	Project management systems	Workflow management systems	Product management systems
Granularity of representation	Medium grained	Coarse-grained	Medium- and fine-grained	Medium-grained
Coverage at the managerial level	Products, activities, resources	Activities, resources	Activities, resources	Products, (activities)
Integration with technical level	Tool integration, document storage	Not supported	Tool integration	Document storage
Support for dynamic design processes	Full support of process evolution	Evolving project plans	Limited (fixed workflows)	Version control for documents
Adaptability	uml models	Not supported	Workflow definitions	Database schema

Project management systems (Kerzner, 1998) such as e.g., Microsoft Project support management functions such as planning, organizing, monitoring, and controlling. The project plan acts as the central document which may be represented in different ways, e.g., as a PERT or GANTT chart. It defines the milestones to be accomplished and provides the foundation for scheduling of resource utilization as well as for cost estimation and control. Project management systems are widely used in practice, but they still suffer from several limitations: project plans are often too coarse-grained, products (documents) are not considered, project plans are not integrated with the actual work performed by engineers, and there is no way to define domain-specific types of project plans.

Workflow management systems (Jablonski and Bußler, 1996 and Lawrence, 1997), e.g., Staffware, FlowMark, or COSA, have been applied in banks, insurance companies, administrations, etc. A workflow management system manages the flow of work between participants, according to a defined procedure consisting of a number of tasks (McCarthy & Bluestein, 1991). It coordinates user and system participants to achieve defined objectives by set deadlines. To this end, tasks and documents are passed from participant to participant in a correct order. Moreover, a workflow management system may offer an interface to invoke a tool on a document either interactively or automatically. Their most important restriction is limited support for the dynamics of design processes. Many workflow management systems assume a statically defined workflow that cannot be changed during execution. In this way, dynamic design processes can be supported only to a limited extent (i.e. the statically known fractions can be handled by the workflow management system). Recently, this problem has been addressed in a few university prototypes (see e.g., Derniame, Baba, and Wastell, 1998 and Georgakopoulos, Prinz, and Wolf, 1999).

In the context of this paper, we use the term *product management system* to refer to all kinds of systems for storing, manipulating, and retrieving the results of design processes. Depending on the context in which they are employed, they are called engineering data management systems, product data management systems (Harris, 1996), software configuration management systems (Tichy, 1994 and Whitgift, 1991), or document management systems. Documentum and Matrix One are examples of such systems which are used in chemical engineering. Documents such as flow diagrams, steady-state and dynamic simulation models, cost estimations, etc. are stored in a database which records the evolution of documents (i.e. their versions) and aggregates them into configurations. In addition, product management systems may offer simple support for the management of activities (e.g., change request processes based on finite state machines), or they may include workflow components, which suffer from the restrictions already discussed above. Their primary focus still lies on the management of products; in particular, management of human resources is hardly considered.

All of the approaches cited above are *domain-independent*. For example, workflow management systems may be applied to arbitrary business processes, and product management systems may be used in different engineering disciplines. We are aware of only a few *domain-specific* approaches which have been developed for chemical engineering. For example, n-dim (Levy et al., 1993; Westerberg et al., 1997) is a distributed and collaborative computer-aided environment for process engineering design; KBDS (Bañares-Alcántara & Lababidi, 1995) deals with the management of design alternatives and design histories. These approaches are better tailored towards design processes in chemical engineering. However, they do not provide comprehensive support for the management of products, activities, and resources. Moreover, they lack the generality of domain-independent systems which can be used in and adapted to different domains.

A MANAGEMENT SYSTEM FOR DE-SIGN PROCESSES

Since current management systems suffer from several limitations explained in the previous section, we have designed and implemented a new management system which addresses these limitations. This system is called *AHEAD* (Jäger, Schleicher, and Westfechtel, 1999b and Westfechtel, 1999). AHEAD is a research prototype that goes beyond commercial systems with respect to the requirements introduced earlier (Table 1):

- *Medium-grained representation.* In contrast to project management systems, design processes are represented at a medium-grained level, allowing managers to effectively control the activities of designers. Management is not performed at the level of milestones, rather, it is concerned with individual tasks such as 'simulate the CSTR reactor'.

- *Coverage and integration at the managerial level.* AHEAD is based on an integrated management model which equally covers products, activities, and resources. In contrast, project and workflow management systems primarily focus on activities and resources, while product management systems are mainly concerned with the products of design processes.

- *Integration between managerial and technical level.* In contrast to project management systems, the AHEAD system also includes support tools for designers that supply them with the documents to work on, and the tools that they may use.

- *Support for the dynamics of design processes.* While many workflow management systems are too inflexible to allow for dynamic changes of workflows during execution, AHEAD supports evolving design processes, allowing for seamless integration of planning, execution, analysis, and monitoring.

- *Adaptability.* Both the structure of management configurations and the operations to manipulate them can be adapted by means of a domain-specific object-oriented model based on the uml (Booch et al., 1999).

Functionality and Concepts

Fig. 8 gives an overview of the AHEAD system. AHEAD offers environments for different kinds of users, which are called modeler, manager, and designer. In the following, we will focus on the functionality that the AHEAD system provides to its users. Its technical realization will be discussed in the next subsection.

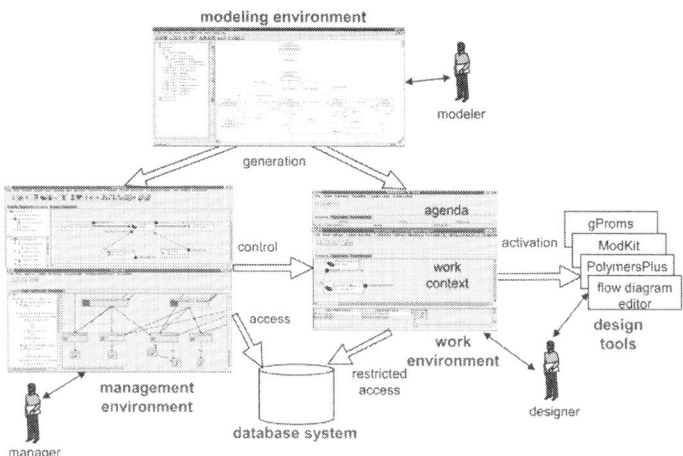

Figure 8: Architecture of the AHEAD system.

The *management environment* supports project managers in planning, analyzing, monitoring, and controlling design processes. It provides graphical tools for operating on management configurations. These tools address the management of activities, products, and resources, respectively (Krapp, Krüppel, Schleicher, & Westfechtel, 1998):

- For *activity management*, AHEAD offers dynamic task nets which allow for seamless interleaving of planning, analyzing, monitoring, and controlling. A task net consists of tasks that are connected by control flow and data flow relationships. Furthermore, feedback in the design process is represented by feedback relationships. Tasks may be decomposed into subtasks, resulting in task hierarchies. The manager constructs task nets with the help of a graphical editor. He may modify task nets at any time while a design process is being executed.

- *Product management* is concerned with documents such as flow diagrams, simulation models, cost estimations, etc. AHEAD offers version control for these documents with the help of version graphs. Relationships (e.g., dependencies) between documents are maintained as well. Versions of documents may be composed into configurations, thereby defining which versions are consistent with each other. The manager may view the version histories and configurations with the help of a graphical tool. In this way, he may keep track of the work results produced by the designers.

- *Resource management* deals with the organizational structure of the enterprise as far as it is relevant to design processes. AHEAD distinguishes between abstract resources (positions or roles) and concrete resources (employees). The manager may define a project team and then assign employees to the project positions.

Management of activities, products, and resources is fully integrated: tasks are assigned to positions, inputs and outputs of tasks refer to document versions. Moreover, AHEAD manages task-specific workspaces of documents and supports invocation of design tools (see below).

AHEAD does not only support managers. In addition, it offers a *work environment* which consists of two major components:

- The *agenda tool* displays the tasks assigned to a designer in a table containing information about state, deadline, expected duration, etc. The designer may perform operations such as starting, suspending, finishing, or aborting a task.

- The *work context tool* manages the documents and tools required for executing a certain task. The designer is supplied with a workspace of versioned documents. He may work on a document by starting a tool such as e.g., a flow diagram editor, a simulation tool, etc.

Please note that the scope of support provided by the work environment is limited. We do not intend to support design activities in detail at a technical level. Rather, the work environment is used to couple technical activities with management. There are other tools which support design activities at a fine-grained level. For example, a process-integrated flow diagram editor (Bayer, Marquardt, Weidenhaupt, and Jarke, 2001) may be activated from the work environment. 'Process-integrated' means that the designer is supported by process fragments which correspond to frequently occurring command sequences. These process fragments encode the design knowledge which is available at the technical level. This goes beyond the scope of the AHEAD system, but it is covered by the overall environment for supporting design processes to which AHEAD belongs as a central component.

Both the management environment and the work environment access a common *management database*. However, they access it in different ways, i.e., they invoke different kinds of functions. The work environment is restricted to those functions which may be invoked by a designer. The management environment provides more comprehensive access to the database. For example, the manager may modify the structure of a task net, which is not allowed for a designer.

Before the AHEAD system may be used to carry out a certain design processes, it must be adapted to the respective application domain Schleicher, 1999. AHEAD consists of a generic kernel which is domain-independent. Due to the generality of the underlying concepts, AHEAD may be applied in different domains such as software, mechanical, or chemical engineering. On the other hand, each domain has its specific constraints on design processes. The *modeling environment* is used to provide AHEAD with domain-specific knowledge, e.g., by defining task types for flow diagram design, steady-state and dynamic simulation, etc. From a domain-specific process model, code is generated for adapting the management and the work environment.

Realization

AHEAD is a research prototype which has been developed to demonstrate novel functionality. To implement AHEAD, we have used powerful homegrown tools for rapid prototyping (see below). This was the fastest way to obtain a demonstrator, which, however, cannot be immediately employed in industry. InSection 6, we will discuss technology transfer, which can be achieved by reusing commercial tools for project, workflow, and product management.

In the current realization of the AHEAD system, *graph technology* plays an important role. All data for representing management configurations are stored in the graph-based database management system GRAS (Kiesel, Schürr, & Westfechtel, 1995). Graph structures and operations are formally specified in the high-level specification language progres (Schürr, Winter, & Zundorf, 1999), which is based on programmed graph transformations. In AHEAD, the specification written in progres describes how management configurations are built up and what operations for manipulating them are offered to end users. From the specification, code is generated which operates on the management database.

For the user interface, we have developed the *UPGRADE* framework (*U*niversal *P*latform for *GRA*ph-Based Application *DE*velopment (Böhlen, Jäger, Schleicher, & Westfechtel, 2002)). UPGRADE is implemented in Java, based on standard libraries and both public-domain and commercial components (ILOG JViews). It mainly focuses on graphical tools, but it also supports e.g., tabular and tree representations. Graphical tools provide external views on the underlying management graph, hiding all of the technical details of the internal representation.

From the work environment, external tools may be started with the help of *wrappers*. Wrappers constitute tool envelopes which are responsible for supplying tools with data (documents checked out from the product management database), preparing the operating system environment (e.g., by setting environment variables), and calling the tools with appropriate parameters. Wrappers are realized on top ofcorba, an infrastructure for distributed object-oriented computing. The realization of wrappers has been performed by another partner participating in the IMPROVE project (Lipperts & Thißen, 1999).

The *modeling environment* is realized with the help of a commercial CASE tool (Rational Rose), which is based on the *uml* (Booch et al., 1999). uml is a language that serves as a standard notation for object-oriented modeling. For the purpose of process modeling, we have adapted and restricted uml according to our requirements (Jäger et al., 1999b). A uml model is automatically transformed into a (part of a) progresspecification (i.e., the end user is not concerned with progres at all). This transformation tool is realized with the help of the OLE interface providing access to Rose's model database.

APPLYING THE MANAGEMENT SYS-TEM TO CHEMICAL ENGINEERING

Within the IMPROVE project, the polyamide6 design process played a key role not only on a conceptual level. In addition, it was used for the development of a demonstrator integrating all software components contributed by the various project partners. The AHEAD system served as the central, coordinating component of this demonstrator (see also Section 2.2). It was integrated with the following tools: thePRIME flow diagram editor (Bayer et al., 2001), ModKit (for creating dynamic simulation models, see (Bogusch, Loehmann, & Marquardt, 2001)), KomPakt (for synchronous multi-media communication (Schüppen et al., 2002)), and Morex (for extruder design) were contributed by partners involved in the IMPROVE project; excel (from Microsoft) and PolymersPlus (from Aspen Tech) are commercial tools which are used for cost calculations and steady-state simulations, respectively. For the demonstrator, a comprehensive demo session was prepared which was successfully presented at the IMPROVE project review in May 2000 and at a national workshop with participants from chemical industry and tool vendors that was held in November 2000.

The Polyamide6 Design Process

Based on the case study introduced in Section 2.2, the polyamide6 design process is represented as a dynamic task net in Fig. 9. This task net was created from a representation in the C3 modeling language which we used in Fig. 2 and Fig. 5. Please note that Fig. 5 shows a very simplified view of the design process. To construct the task net of Fig. 9, we used a more detailed C3 model.

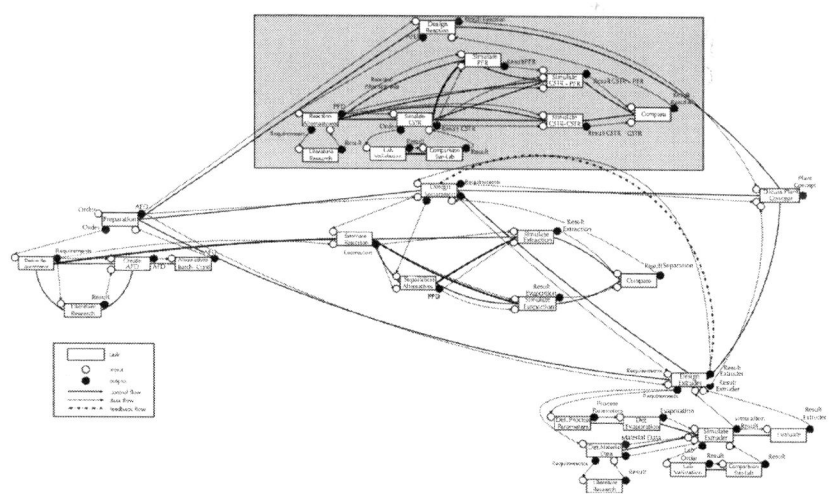

Figure 9: Task net for the polyamide6 design process.

As we will demonstrate in the demo session of Section 5.2, the task net of Fig. 9 is not available at the start of the design process. Rather, it is built up as the design proceeds. In an early phase, the task net will just contain a top-level decomposition into design tasks for reaction, separation, and compounding, respectively. However, in the IMPROVE project we were dealing with a case study which was designed beforehand (and was used to plan the development of the demonstrator). Therefore, a C3 model of the overall design process was already available in this artifical setting; it would not be available in real-world use.

The mapping from C3 to task nets was performed manually, albeit in a systematic manner. In general, human expertise is required to perform the mapping. On the other hand, there is a straightforward initial mapping which may be obtained by applying the following simple rules:

- Each activity is mapped onto a task. This also applies to cooperative activities, which occur only once in a task net.[5]

- Ordering relationships between activities are mapped onto control flows. While forks and joins are represented in C3 by bars, they are not shown explicitly in task nets. Rather, they appear as bundles of outgoing and incoming control flows, respectively.

- An arrow from an activity to a document is mapped onto an output parameter of the respective task. For each arrow from a document to an activity, an input parameter is generated, and output and input are connected by a data flow.

Human expertise is required e.g., for introducing task hierarchies. The C3 model which we used as input was flat. Based on domain knowledge, reasonable subprocesses may be identified. In our case study, we have introduced subprocesses for designing the reaction, the separation, and the compounding of the chemical process. In addition, there is a preparation phase in which the overall design problem is decomposed, and a final integration phase, where the overall plant design is synthesized and discussed among the involved design experts.

Please note that reaction, separation, and compounding are designed in an integrated way:

- To design the separation, input is required with respect to incoming streams. In the sample process, this input is generated by an initial estimation which is refined later on. In this way, reaction and separation may be studied in parallel. This speeds up the overall design process.

- With respect to separation and compounding, there exist some degrees of freedom with respect to the decomposition of the overall chemical process. In particular, if separation is performed with the help of evaporation, evaporation can be performed partly in the extruder. Therefore, the respective design processes are arranged in a feedback loop for optimizing the chemical process.

In the demo session to be presented below, we focus on the design of the reaction (shaded region in Fig. 9). After an initial PFD has been created which contains multiple design variants, each of these variants is explored by means of simulations and (if required) laboratory experiments. In a final step, these alternatives are compared against each other, and the most appropriate one is selected. This simplified part of the design process suffices to demonstrate many of the essential features provided by the AHEAD system; furthermore, it is sufficiently small to be presented in this paper.

Demo Session

In this subsection, we illustrate the functionality of the AHEAD system with the help of some snapshots. We will primarily focus on the management environment; the work environment will be discussed rather briefly. The modeling environment goes beyond the scope of this paper; see Jäger et al. (1999a).

Fig. 10 presents a snapshot from the management environment taken in an early stage of the polyamide6 design process. The upper region on the left displays a tree view of the task hierarchy. The lower left region offers a view onto the resources available for task assignments (see also Fig. 11). A part of the overall task net is shown in the graph view on the right-hand side. Each task is represented by a rectangle containing its name, the position to which the task has been assigned, and an icon representing its state (e.g., the gear-wheels represent the state Active, and the hour-glass stands for the state Waiting). Black and white circles represent outputs and inputs, respectively. These are connected by data flows (thin arrows). Furthermore, the ordering of task execution is constrained by control flows (thick arrows). Hierarchical task relations (decompositions) are represented by the graphical placement of the task boxes (from top to bottom) rather than by drawing arrows (which would clutter the diagram).

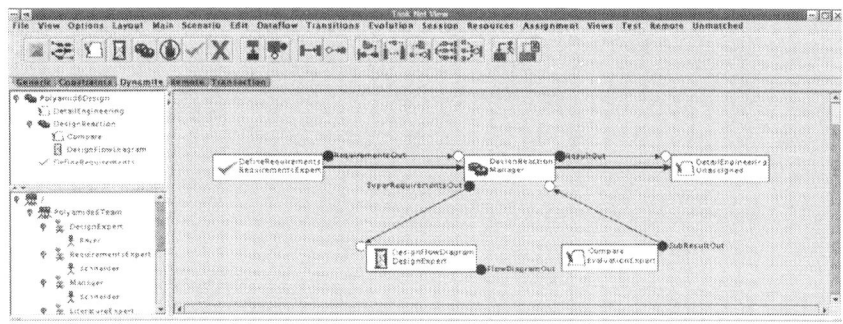

Figure 10: Initial task net (management environment).

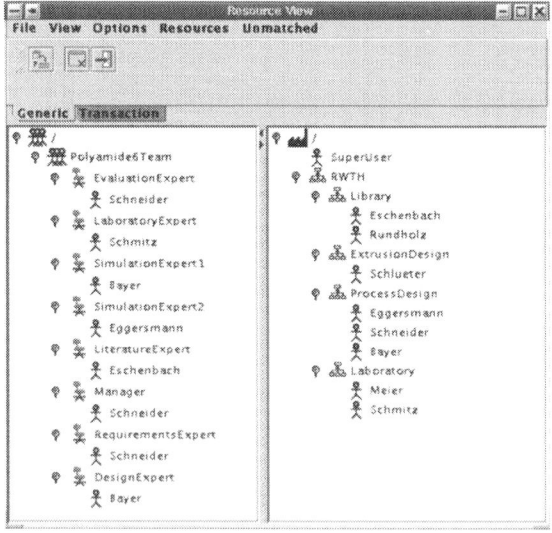

Figure 11: Resource view (management environment).

Please recall that the demo session deals only with the reaction part, i.e., we do not consider separation, extrusion, etc. On the top level of the task net of Fig. 10, the process is decomposed into three tasks: Define Requirements, Design Reaction, and Detail Engineering. Only the design of the reaction is elaborated in the sequel. In this early stage, it is only known that initially some reaction alternatives have to be designed. The result of this design task is documented in a flow diagram. Furthermore, at the end these alternatives have to be compared, and a decision has to be performed. All other tasks—e.g., for performing simulations and laboratory experiments—have to be filled in later. Thus, the initial task net is incomplete.

In addition to the initial task net, the manager has also used the resource management tool for building up his project team (Fig. 11). The region on the left displays the structure of the polyamide6 design team. Each position (represented by a chair icon) is assigned to a team member. Analogously, the region on the right shows the departments of the company. From these departments, the team members for a specific project are taken for a limited time span. The management environment offers commands for defining both the team and the department structure, and for assigning persons to positions in design teams. Please note that tasks are assigned to positions rather than to actual employees (see lower left view in Fig. 10). In this way, assignment is decomposed into two steps. The manager may assign a task to a certain position even if this position has not been filled yet. Moreover, if a different employee is assigned to a position, the task assignments need not be changed: The tasks will be redirected to the new employee automatically.

The work environment is illustrated in Fig. 12. As a first step, the user logs into the system (not shown in the figure). After that, AHEAD displays an agenda of tasks assigned to this user (more precisely: assigned to the roles played by this user). In Fig. 12, the agenda is shown in the top window. Since the user Bayer plays the role of the design expert, the agenda contains the task Design Flow Diagram. After the user has selected a task from the agenda, the work context for this task is opened (bottom window). The work context graphically represents the task, its inputs and outputs, as well as its context in the task net (here, the context includes the parent task which defines the requirements to the flow diagram to be designed). Furthermore, it displays a list of all documents needed for executing this task. For some selected document, the version history is shown on the right (so far, there is only one version of the requirements definition which acts as input for the current task).

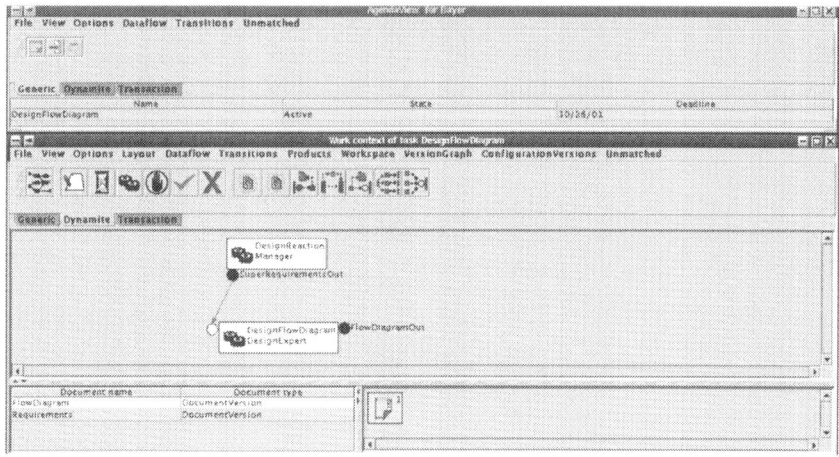

Figure 12: Work environment.

From the work context window, the user may activate design tools for operating on the documents contained in the workspace. Here, the user invokes a flow diagram editor (Bayer et al., 2001) in order to insert reaction alternatives into the flow diagram for the polyamide6 process. The flow diagram editor, which was also developed in the IMPROVE project, is based on MS Visio, a commercial drawing tool, which was integrated with the PRIME process engine prime (Pohl et al., 1999). The flow diagram editor supports hierarchical flow diagrams (abstract or process flow diagrams); furthermore, it may represent alternative refinements (*variants*) for blocks occurring in the flow diagram. The flow diagram editor offers fine-grained process fragments which are used for guidance and automation. These fragments incorporate domain-specific knowledge. For example, there is a process fragment which assists the designer in introducing reaction alternatives into the flow diagram (based on experience from previous design processes).

The resulting flow diagram is displayed in Fig. 13. The chemical process is decomposed into reaction, separation, and compounding. The reaction is refined into four variants. For our demo session, we assume that initially only two variants are investigated (namely a single CSTR and PFR, respectively). That is, at the current state of design the alternatives on the right-hand side have not yet been introduced into the flow diagram; they will be considered later on.

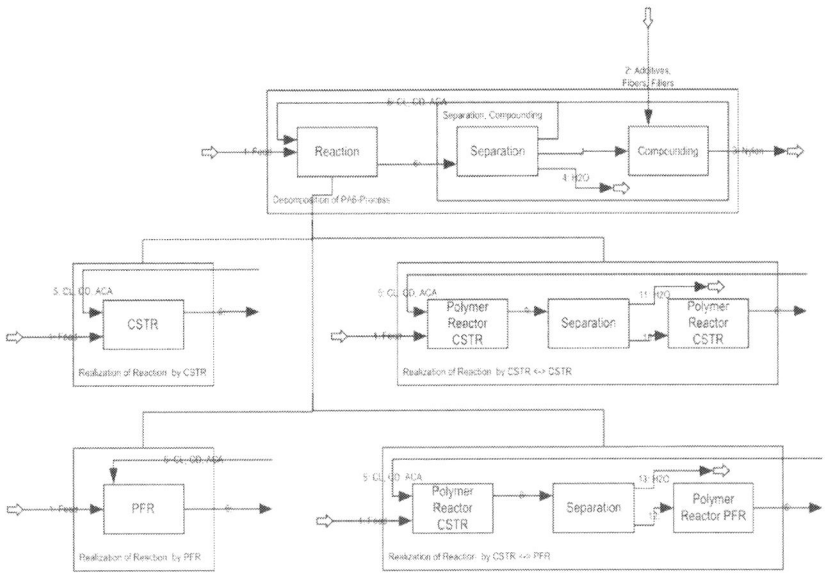

Figure 13: Reaction alternatives in the PFD.

After the generation of the four variants, the manager extends the task net with tasks for investigating the alternatives that have been introduced so far (*product-dependent task net*, Fig. 14). Please note the control flow relation between the new tasks: The manager has decided that the CSTR should be investigated first so that experience from this alternative may be re-used when investigating the PFR. Furthermore, we would like to emphasize that the design task is not terminated yet. As to be demonstrated below, the designer waits for feedback from simulations in order to enrich the flow diagram with simulation data. Depending on these data, it may be necessary to investigate further alternatives.

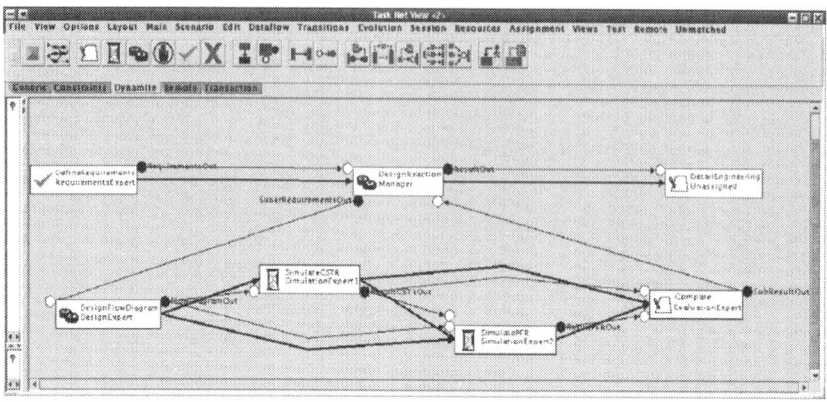

Figure 14: Extended tesk net (management environment).

Subsequently, the simulation expert creates a simulation model (using PolymersPlus) for the CSTR reactor and runs the corresponding simulations. The simulation results are validated with the help of laboratory experiments. After these investigations have been completed, the flow diagram can be enriched with simulation data such as flow rates, pressures, temperatures, etc. To this end, a *feedback flow*— represented by a dashed arrow—is inserted into the task net (Fig. 15). The feedback flow is refined by a data flow, along which the simulation data are propagated. Then, the simulation data are introduced into the flow diagram.

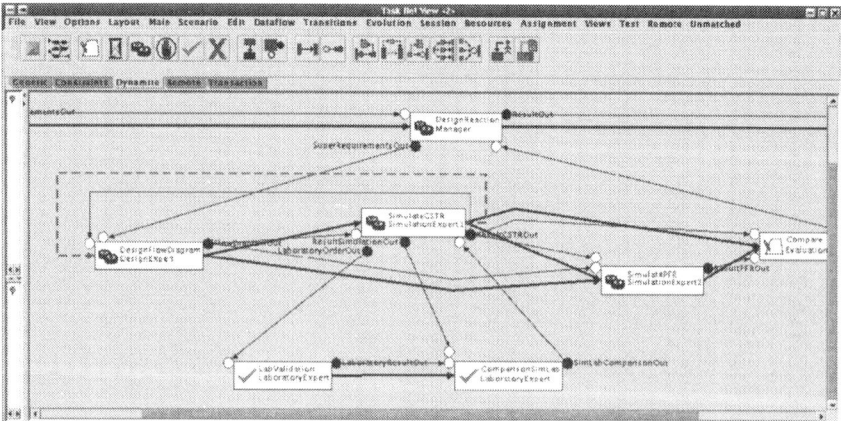

Figure 15: Feedback and simultaneous engineering (management environment).

Please note that the tasks for designing the flow diagram and for investigating the CSTR are active at the same time. The semantics of control flows is defined such that tasks connected by control flows can be active simultaneously. In this way, we support *simultaneous engineering* (Bullinger & Warschat, 1996). As a consequence, we cannot assume that the work context of a task is stable with respect to its inputs. Rather, a predecessor task may deliver a new version that is relevant for its successors. This is taken care of by a sophisticated release policy built into the model underlying dynamic task nets (Westfechtel, 1999).

After the alternatives CSTR and PFR have been elaborated, the evaluation expert compares all explored design alternatives. Since none of them performs satisfactorily, feedback is raised to the design task. Note this is an example of far-reaching feedback (from the end to the start of the subprocess for designing the reaction part). Here, we assume that the designer has already terminated the design task. As a consequence, the design task has to be reactivated. Reactivation is handled by creating a new *task version*, which may or may not be assigned to the same designer as before. New design alternatives are created, namely a CSTR–CSTR and a CSTR–PFR cascade, respectively (see again Fig. 13). Furthermore, the task net is augmented with corresponding simulation tasks (Fig. 16 [6]). After that, the new simulation tasks are delegated to simulation experts, and simulations are carried out accordingly. Eventually, the most suitable reactor alternative is selected.

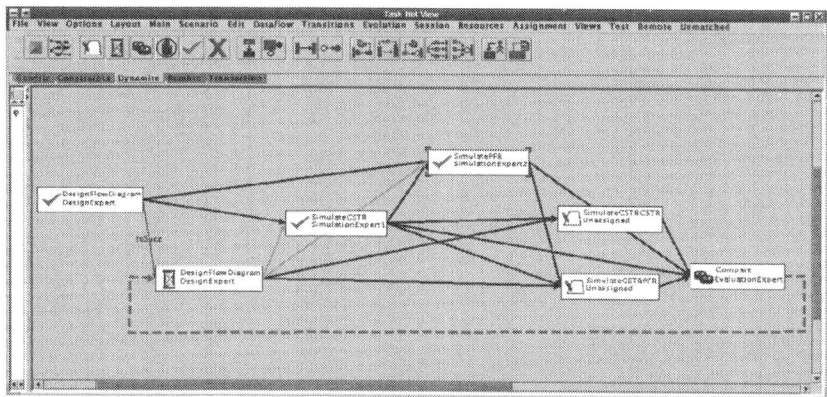

Figure 16: Far-reaching feedback (management environment).

So far, we have primarily considered the management of activities. Management of products, however, is covered as well. This is illustrated by the snapshot in Fig. 17, which is again taken from the management environment. It shows a tree view on the products of design processes on the left and a graph view on the right. Products are arranged into *workspaces* that are organized according to the task hierarchy. Workspaces contain sets of *versioned documents*.

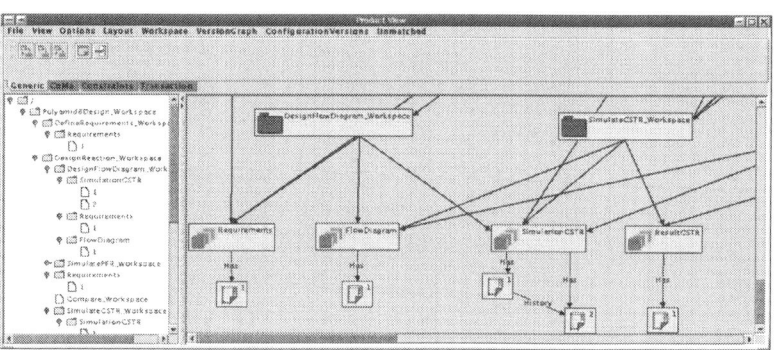

Figure 17: Product view (management environment).

Generally speaking, a *version* represents some state (or snapshot) of an evolving document. We distinguish between *revisions*, which denote temporal versions, and *variants*, which exist concurrently as alternative solutions to some design problem. Revisions are organized into sequences, variants result in branches. Versions are connected by *history relationships*. In Fig. 17, there is a history relationship between revisions 1 and 2 of Simulation CSTR, the simulation model for the CSTR reactor. In general, the version history of a document (flow diagram, simulation model, etc.) may evolve into an acyclic graph (not shown in the snapshot).

Please note that in Fig. 17 there is only one version of the flow diagram. Here, we rely on the capabilities of the flow diagram editor to represent multiple variants. Still, the flow diagram could evolve into multiple versions at the managerial level (e.g., to record snapshots at different times). Moreover, in the case of a flow diagram editor with more limited capabilities (no variants), variants would be represented at the managerial level as parallel branches in the version graph.

TECHNOLOGY TRANSFER

In this section we discuss how the obtained *results* can be *transferred* to industry in order to improve the state of the art of coordinating industrial design processes.

Ways of Technology Transfer

Let us start with discussing the different *ways* of *technology transfer* of the obtained results. Transfer ranges from evaluating the new concepts introduced above to implementing the functionality of AHEAD in an industrial context by using existing systems:

- *Conceptual transfer* (transfer 1): Conceptual transfer includes to see whether the concepts can be explained and are accepted, whether they solve real-world problems, and whether the extension of industrial practice can be managed. This can be achieved by demonstrating AHEAD using a prepared demo session as described in the previous section, by discussing the underlying concepts with industrial partners, etc.

- *Evaluation of the demonstrator prototype* (transfer 2): Here, we plan the use of the AHEAD demonstrator for a small industrial project, again in a project together with an industrial partner. This requires that control on a medium-grained level is available in the company and there is the intention to manage that level. Discussions with different industrial partners, where the AHEAD system was demonstrated, have shown that the necessity of control on medium-grained level is seen. An exemplary use of the system also includes the evaluation of the user interfaces for the work environment, the management environment, and the modeling environment.

- *Realization of an industrial system with comparable functionality* (transfer 3): Comparable functionality here means that the requirements stated in Section 3 and Section 4 are met, or are at least partially met. So, we aim at getting as much functionality of the AHEAD system as possible or, at least, we do not loose essential functionality. Thereby, we take existing systems and integrate them on a bottom–up basis.

Different Strategies and Common Problems

In the rest of this section we *concentrate* on *realization transfer* (transfer 3). There are *two ways* how an industrial system (with the functionalities of AHEAD) can look like.

The AHEAD system itself is acting as a *management integration instance*. Existing industrial systems are wrapped and incorporated into the integrated industrial system (Fig. 18).

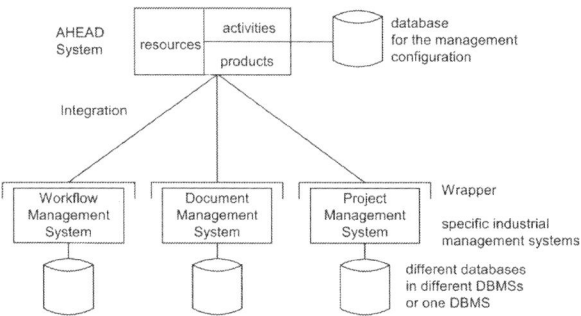

Figure 18: The AHEAD system as integrating instance of commercial systems.

We take again industrial systems for different purposes. On top of them a *new system* is realized which has essential AHEAD functionality. However, this system realization is using *industrial components* as well as a *conventional* industrial software development *process*, i.e., no rapid prototyping (Fig. 19).

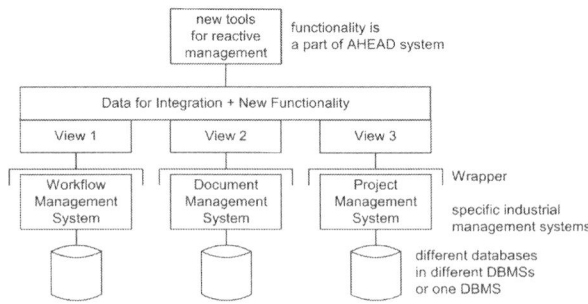

Figure 19: Realization of AHEAD functionality using industrial infrastructure and systems.

For the rest of this subsection, we give some remarks and state some implementation problems which hold for both solutions (a) and (b). The reader should note that both ways *follow* the overall *bottom–up approach*of IMPROVE as explained in Section 1 (Fig. 1) inasmuch as existing tools are re-used.

Industrial *systems* to be used in a bottom–up approach have to be *stripped* in order to concentrate on their main functionality (separation of concerns). For example, a document management system is only used for the management of documents and not for managing activities and resources, although such systems may include some restricted functionality for that purpose. Otherwise, we would have to handle activity management in different industrial systems and in the integrated system on top as well. It is difficult enough to handle it both on the level of existing systems and on the integration instance level, where different existing systems are unified and integrated.

Both variants (a) and (b) introduce *wrappers* for existing systems. Wrappers have two aspects. First, we need a *functional interface* which is coarse- (start/end) or fine-grained (invoking all commands) depending on the openness of existing systems. Second, there is a *data interface* in the form of a view. So, we do not access the mostly cryptic internal data of an industrial application. Instead, we define a data abstraction layer by which we get rid of data structuring details. Of course, the data views of industrial systems have to be homogeneous with respect to granularity and structure.

In both variants (a) and (b) we internally have to maintain and to retrieve the data of the *management configuration*. In variant (a), the management configuration is represented in the AHEAD system. In variant (b), the management configuration has to be built up by using views on existing systems. In addition, the mutual relations between these parts have to be defined (e.g., between activities and resources) in the second variant.

Discussion of Both Implementation Alternatives

We now discuss the two ways of realizing an industrial system in more detail.

Way (a) can again be split into two variants. The first variant (a1) is to have the *full description* for products, activities, and resources within the management configuration of AHEAD. The second variant (a2) is that the AHEAD system only contains a *filtered* portion of the *data* of the existing systems *together with* some*integration* data. Then, the details are stored in the industrial subsystems.

In both cases (a1) and (a2) we find the *semantic models* of design processes in AHEAD at a *central place*, guaranteeing that the semantic submodels fit together. Within industrial systems a part of these models cannot be mapped (for example the integration of the three perspectives for products, activities, and resources). In solution (a1), we find the *detail information* for the submodels in the industrial systems as well as in AHEAD. As the access is simpler *via AHEAD*, we make data access and maintenance using this system. In solution (a2), we find only a part of the information in AHEAD. So, access of integrating information as well as coarse information is done via AHEAD, whereas the details are accessed *using the*view interfaces* for industrial systems.

Let us discuss the *coupling* of *AHEAD* and the *industrial systems* by taking the task part of AHEAD and a workflow system as an example. As discussed in Section 3, workflow systems which allow to structure a task plan either do not regard dynamic aspects at all (static plans) or they allow subnet calling thereby distributing the static structure of a dynamically changing net over subnet definitions and the run time storage of the executing system. In solution (a1) the complete net structure is built up in AHEAD, the given workflow system is only used for executing the overall net or subnets. In solution (a2) only the global structure is kept within AHEAD whereas the detailed subnet structures are found in the workflow system. Thereby, execution is performed on a coarse level in AHEAD and on a more detailed level in the workflow system.

Let us now shortly *discuss alternative (b)* of above. Here, the AHEAD functionality of semantic submodels for products, activities, and resources, integration of submodels, interleaved invocation of structuring, analyzing and execution, adaptation mechanisms, etc. have to be re-implemented in the new industrial system on top of existing systems using only conventional software development techniques. This is not trivial. However, the essential features can be realized with some (remarkable) effort.

Comparing both solutions (a) and (b) we can state: Way (a) is much simpler, as the AHEAD system with its functionality is a part of the integration solution. However, the AHEAD system is just a demonstrator from academia and not an industrial system. So way (b) is more likely to be used. However, then the functionality of AHEAD has to be re-implemented, a nontrivial software development task.

As a consequence of the above discussion, the question now arises why we did not realize the AHEAD system by a bottom–up approach from the very beginning. The answer to this question is that we have firstly looked for the right concepts. In order to find them, the different possibilities had to be investigated. For playing around, a realization machinery as explained in Section 4.2 is much more convenient than coding different variants. Now, as a demonstrator is available, and if it has successfully passed the different stages of proof of concept (transfer 1 and 2), we are tackling the question of how an industrial system can look like. Our research is still continuing using the implementation machinery of Section 4.2 in order to quickly find further concepts/ mechanisms which, again in a second run, are transferred to industry. Therefore, also in the future, we *distinguish* between *scientific investigation* on the one hand and *transfer projects* on the other.

CONCLUSION AND OUTLOOK

We have presented the AHEAD system for managing design processes in chemical engineering. AHEAD goes beyond the functionality of commercial systems in various respects. Management is performed on a medium-grained level, allowing to effectively coordinate the work of designers. Products, activities, and resources are equally taken into account and are managed in an integrated way with the help of the management environment. Through its work environment, AHEAD is coupled with external tools supporting designers in their technical work. By seamless interleaving of planning and execution, AHEAD takes dynamic design processes into account. Finally, AHEAD may be adapted to different application domains with the help of its modeling environment.

Ongoing and future work addresses the following areas:

- *Technology transfer*: As discussed in Section 6, current work addresses the evolution of the AHEAD prototype into a system which may be used in an industrial context.

- *Improved functionality*: We are still extending the AHEAD prototype to improve its capabilities in various respects, e.g., concerning the management of distributed design processes (Becker, Jäger, Schleicher, & Westfechtel, 2001) or even more sophisticated support for process evolution (Schleicher, 2002).

- *Synergy*: We are integrating AHEAD more tightly with other components of the integrated design environment developed in the IMPROVE project. For example, we are designing an integration tool for coupling flow diagrams and task nets (so far, this coupling has to be performed manually), and we are investigating potentials for fine-grained process support for managers using the AHEAD system.

ACKNOWLEDGEMENTS

We are indebted to all researchers who have worked in the IMPROVE project and have contributed to the AHEAD system in one way or the other. In particular, we are indebted to A. Schleicher and D. Jäger, who have made essential contributions in their Ph.D. theses, and to B. Bayer, M. Eggersmann, and W. Marquardt.

REFERENCES

1. Ban˜ares-Alca´ntara, R. (1995). Design support systems for process engineering 1. Requirements and proposed solutions for a design process representation. Computers and Chemical Engineering 19 (3), 267-/277.

2. Ban˜ares-Alca´ntara, R., & Lababidi, H. (1995). Design support systems for process engineering-II. KBDS: An experimental prototype. Computers and Chemical Engineering 19 (3), 279-/301.

3. Bayer, B., Eggersmann, M., Gani, R., & Schneider, R., (2002). Case studies in process design. In: Braunschweig & Gani, 2002.

4. Bayer, B., Marquardt, W., Weidenhaupt, K., & Jarke, M., 2001. A flowsheet centered architecture for conceptual design. In: Gani, R., & Jorgensen, S. (Eds.), European symposium on computer aided process engineering 11 (pp. 345-/350), Elsevier.

5. Becker, S., Haase, T., Westfechtel, B., & Wilhelms, J., 2002. Integration tools supporting cooperative development processes in chemical engineering. In: Ehrig, H., Kra¨mer, B.J., & Ertas, A. (Eds.), Proceedings of the sixth biennial world conference on integrated design and process technology (IDPT2002). Society for Design and Process Science, Pasadena, California.

6. Becker, S., Ja¨ger, D., Schleicher, A., & Westfechtel, B., 2001. A delegation-based model for distributed software process management. In: Ambriola, V. (Ed.), Proceedings eighth european workshop on software process technology (EWSPT 2001). LNCS 2077 (pp. 130-/144). Witten, Germany: Springer.

7. Biegler, L. T., Grossmann, I. E., & Westerberg, A. W. (1997). Systematic Methods of Chemical Process Design. Upper Saddle River: Prentice Hall. Blass, E. (1997). Entwicklung verfahrenstechnischer Prozesse: Methoden, Zielsuche, Lo¨sungssuche, Lo¨sungsauswahl. Berlin, Germany: Springer.

8. Bogusch, R., Loehmann, B., & Marquardt, W. (2001). Computeraided process modeling with ModKit. Computers and Chemical Engineering 25 (7-/8), 963-/995.

9. Bo¨hlen, B., Ja¨ger, D., Schleicher, A., & Westfechtel, B., 2002. UPGRADE: Building interactive tools for visual languages. In: Callaos, N., Hernandez-Encinas, L., & Yetim, F. (Eds.), Vol. I: Information Systems Development I. Proceedings of the sixth world multiconference on systemics, cybernetics, and informatics (SCI 2002) (pp. 17-/22). Orlando, Florida.

10. Booch, G., Rumbaugh, J., & Jacobson, I. (1999). The Unified Modeling Language User Guide . Reading, Massachusetts: Addison Wesley.

11. Braunschweig, B., & Gani, R. (Eds.). Software Architectures and Tools for Computer Aided Process Engineering. Elsevier Publishers, Vol. 11.

12. Bullinger, H.-J., Warschat, J., Concurrent Simultaneous Engineering Systems (1996). Berlin, Germany: Springer.Springer, Berlin, Germany Derniame, J.-C., Baba, A.K., Wastell, D., Software Process: Principles, Methodology, and Technology. LNCS 1500 (1998). Berlin, Germany: Springer.

13. Deutsches Patentamt, 1969. Verfahren and Vorrichtung zum kontinuierlichen Polykondensieren von Lactamen. Patent number P 14 95 198.5 (B78577).

14. Douglas, J. M. (1988). Conceptual Design of Chemical Processes. New York: McGraw-Hill.

15. Eggersmann, M., Hackenberg, J., Marquardt, W., & Cameron, L., 2002. Applications of modeling*/a case study from process design . In: Braunschweig & Gani, 2002.

16. Foltz, C., Killich, S., Wolf, M., Schmidt, L., & Luczak, H. (2001). Task and information modeling for cooperative work. In: M. Smith, & G. Salvendy (Eds.), Systems, Social and Internationalisation Design Aspects of HumanComputer Interaction. Volume 2 of the Proceedings of HCI International 2001. Lawrence Erlbaum Associates, Mahwah, New Jersey (pp. 172-/176).

17. Georgakopoulos, D., Prinz, W., & Wolf, A.L. (Eds.), 1999. Proceedings of the International Joint Conference on Work Activities Coordination and Collaboration (WACC-99). Vol. 24-/2 of ACM SIGSOFT Software Engineering Notes. San Francisco, CA: ACM Press.

18. Gerrens, H. (1981). On selection of polymerization reactors. German Chemical Engineering 4 , 1-/13.

19. Harris, S.B. (1996). Business strategy and the role of engineering product data management: A literature review and summary of the emerging research questions. Proceedings of the Institution of Mechanical Engineers, Part B (Journal of Engineering Manufacture) 210(B3) 207-/220.

20. Jablonski, S., & Bußler, C. (1996). Workflow Management-Modeling Concepts and Architecture . Bonn, Germany: International Thomson Publishing.

21. Ja¨ger, D., Schleicher, A., & Westfechtel, B. (1999). AHEAD: A graphbased system for modeling and managing development processes. In: Nagel, et al. (pp. 325-/339).

22. Ja¨ger, D., Schleicher, A., & Westfechtel, B. (1999b). Using UML for software process modeling. In O Nierstrasz & M. Lemoine (Eds.), Software engineering-ESEC/FSE '99. LNCS 1687 (pp. 91-/108). Toulouse, France: Springer.

23. Kerzner, H. (1998). Project Management: A Systems Approach to Planning, Scheduling, and Controlling. New York: Wiley.

24. Kiesel, N., Schu¨rr, A., & Westfechtel, B. (1995). GRAS, a graphoriented software engineering database system. Information Systems 20 (1), 21-/51.

25. Kohan, M., Nylon Plastics Handbook (1995). Munich, Germany: Carl Hanser Verlag.

26. Krapp, C.-A., Kru¨ppel, S., Schleicher, A., & Westfechtel, B. (1998). Graph-based models for managing development processes, resources, and products. In: G. Engels & G. Rozenberg, TACT '98- 6th International Workshop on Theory and Application of Graph

27. Transformation. LNCS 1764, pp. 455-/474. Paderborn, Germany: Springer.

28. Lawrence, P., Workflow Handbook (1997). Chichester, UK: Wiley.

29. Levy, S., Subrahmanian, E., Konda, S., Coyne, R., Westerberg, A., Reich, Y., 1993. An overview of the n-dim environment. Tech. Rep. EDRC-05-65-93, Carnegie Mellon University, Pittsburgh, Pennsylvania.

30. Lipperts S., Thißen D., 1999, CORBA wrappers for a-posteriori management: An approach to integrating management with existing heterogeneous systems. In: Proceedings third International Conference on Formal Methods for Open Object-Based Distributed Systems. Florence, Italy, pp. 273-/280.

31. Marquardt, W., & Nagl, M., 1999. Tool integration via interface standardization? In: Computer Application in Process and Plant Engineering-Papers of the 36th Tutzing Symposion. Vol. 135 of DECHEMA Monographie. Wiley VCH, Weinheim, Germany, pp. 95-/126.

32. McCarthy, J., & Bluestein, W., 1991. The Computing Strategy Report: Workflow's Progress. Forrester Research, Inc.

33. McGuire, M. L., & Jones, K. (1989). Maximizing the protential of process engineering databases. Chemical Engineering Progress 85 (11), 78-/83.

34. Mostow, J. (1985). Toward better models of the design process. The AI Magazine 6 (1), 44-/57.

35. Nagl, M., Schneider, R., & Westfechtel, B., Synergetische Verschra"nkung bei der A-posteriori-Integration von Werkzeugen. In: Nagl, M., & Westfechtel, B. (Eds.), Modelle, Werkzeuge and Infrastrukturen zur Unterstu"tzung von Entwicklungsprozessen. Wiley VCH, Weinheim, Germany, in press.

36. Nagl, M., & Westfechtel, B., Integration von Entwicklungssystemen in Ingenieuranwendungen (1998). Heidelberg, Germany: Springer. PIEBASE Working Group 2, 1998. Activity Model. Available from http://cic.nist.gov/piebase (Accessed 8 October, 2001).

37. Pohl, K., Klamma, R., Weidenhaupt, K., Do"mges, R., Haumer, P., & Jarke, M. (1999). Process-integrated (modelling) environments (PRIME): Foundations and implementation framework. ACM Transactions on Software Engineering and Methodology 8 (4), 343-/ 410.

38. Ponton, J. (1995). Process systems engineering: Halfway through the first century. Chemical Engineering Science 50 (24), 4045-/4059.

39. Rudd, D., Powers, G., & Siirola, J. (1973). Process Synthesis. Englewood Cliffs: Prentice-Hall.

40. Schleicher, A. (1999). In: Nagl, et al., Formalizing UML-based process models using graph transformations. pp. 341-/358.

41. Schleicher, A., 2002. Roundtrip process evolution support in a wide spectrum process management system. Ph.D. thesis, Aachen University of Technology, Aachen, Germany.

42. Schu"ppen, A., Trossen, D., & Wallbaum, M. (2002). Shared workspace for collaborative enginering. Annals of Cases on Information Technology IV, 119-/130.

43. Schu"rr, A., Winter, A., & Zu"ndorf, A. (1999). The PROGRES approach: Language and environment. In: H. Ehrig, G. Engels, H.-J. Kreowski & G. Rozen-berg. Handbook on Graph Grammars and Computing by Graph Transformation: Applications, Languages,

and Tools, vol. 2: Applications, Languages and Tools. (pp. 487-/550). Singapore: World Scientific.

44. Smithers, T., & Troxell, W. (1990). Design is intelligent behaviour, but what's the formalism? Artificial Intelligence for Engineering Design. Analysis and Manufacturing 4 (2), 89-/98.

45. Thayer, R.H. (1988). Software engineering project management: A top-/down view. In: R.H. Thayer. Tutorial: Software Engineering Project Management, (pp. 15-/54). Washington, DC: IEEE Computer Society Press.

46. Tichy, W.F. Configuration Management of Trends in Software , vol. 2 (1994). New York: Wiley.

47. Westerberg, A. W., Subrahmanian, E., Reich, Y., & Konda, S. (1997). Designing the process design process. Computers & Chemical Engineering 21 (S), S1-/S9.

48. Westfechtel, B. (1999). Models and Tools for Managing Development Processes. LNCS 1646. Heidelberg, Germany: Springer.

49. Whitgift, D. (1991). Methods and Tools for Software Configuration Management. Wiley Series in Software Engineering Practice . New York: Wiley.

50. Yin, R. (1984). Case Study Research. Beverly Hills: Sage Publications.

Scope for Industrial Applications of Production Scheduling Models and Solution Methods

Iiro Harjunkoski[a], Christos T. Maravelias[b],
Peter Bongers[c, j], Pedro M. Castro[d], Sebastian
Engell[e],Ignacio E. Grossmann[f], John Hooker[g],
Carlos Méndez[h], Guido Sand[a], and John Wassick[i]

[a]ABB Corporate Research, Wallstadter Str. 59, 68526 Ladenburg, Germany

[b]University of Wisconsin, Madison, WI 53706, United States

[c]Unilever R&D Vlaardingen, 3133 AT Vlaardingen, The Netherlands

[d]Laboratório Nacional de Energia

[e]Geologia (LNEG), 1649-038 Lisboa, Portugal e Department of Biochemical and Chemical Engineering, Technische Universität Dortmund, 44221 Dortmund, Germany

[f]Department of Chemical Engineering, Carnegie Mellon University, Pittsburgh, PA 15213, United States

[g]Tepper School of Business, Carnegie Mellon University, Pittsburgh, PA 15213, United States

[h]INTEC (UNL – CONICET), Güemes 3450, 3000 Santa Fe, Argentina

ᶦThe Dow Chemical Company, United States j Eindhoven University of Technology, 5612 AZ Eindhoven, The Netherlands

ABSTRACT

This paper gives a review on existing scheduling methodologies developed for process industries. Above all, the aim of the paper is to focus on the industrial aspects of scheduling and discuss the main characteristics, including strengths and weaknesses of the presented approaches. It is claimed that optimization tools of today can effectively support the plant level production. However there is still clear potential for improvements, especially in transferring academic results into industry. For instance, usability, interfacing and integration are some aspects discussed in the paper. After the introduction and problem classification, the paper discusses some lessons learned from industry, provides an overview of models and methods and concludes with general guidelines and examples on the modeling and solution of industrial problems.

INTRODUCTION AND MOTIVATION

Scheduling is a decision-making process that plays an important role in most manufacturing and service industries (Pinedo & Chao, 1999). Scheduling problems arise in almost any type of industrial production facilities (Pulp and Paper, Metals, Oil and Gas, Chemicals, Food and Beverages, Pharmaceuticals, Transportation, Service, Military, etc.) where given tasks need to be processed on specified resources. For instance, in a chemical process, the production must be appropriately planned to ensure that the equipment, material, utilities, personnel and other resources are available at the plant when they are needed to realize the production tasks. Production scheduling comprises the activity of planning the production of e.g. customer orders in detail on a given production facility. In a nutshell, it commonly boils down to the following main decisions:

- What tasks to execute?
- Where to process the production tasks (assignment of tasks to resources)?

- In which sequence to produce (sequencing of tasks)?
- When to execute the production tasks (timing of tasks)?

Furthermore, if not already done by the enterprise resource planning (ERP) systems, the customer orders need to be distributed into production orders (batching). In some cases, only parts of these decisions need to be made, however often they are all relevant to the production. While they are typically also strongly coupled through the synchronization of the resource utilization, ideally the decisions should be taken simultaneously.

Traditionally, production scheduling or short-term planning has been done manually by trained individuals using pen and paper, planning cards, e.g. Kanban or spreadsheets. In the course of time, many companies have identified and documented best operating practices for meeting their specific production targets. Along increasing production volumes, larger product portfolio, alternative production recipes and – especially more recently – volatile customer orders and high pressure to save on production and energy costs, manual scheduling has become extremely challenging. Due to this complexity and required flexibility it is very difficult to ensure a profitable production without any optimization support. A good optimization solution can result in significant savings through better capacity utilization. Apart from economic benefits, good schedules can also contribute to reducing the environmental load, energy demand, violations of various regulations and help coping more efficiently with uncertainties, both in production as well as in customer order levels.

A number of review papers on scheduling have been written across different scientific communities, e.g.Floudas and Lin (2004), Méndez, Cerdá, Grossmann, Harjunkoski, and Fahl (2006), Li and Ierapetritou, 2008a and Li and Ierapetritou, 2008b, Ribas, Leisten, and Framiñan (2010), Phanden, Jain, and Verma (2011), and Maravelias (2012). Due to the wide range of scheduling problems, a number of approaches have been developed to make production scheduling easier and yield better solutions. Some of the approaches comprise computer-supported manual scheduling (e.g. interactive Gantt charts), expert systems (imitating the human behavior), mathematical programming (LP, MILP and MINLP), various heuristics, evolutionary algorithms and different artificial intelligence (AI)-methods. Recently, also approaches trying to tackle uncertainty have been introduced (stochastic optimization).

Most of the above methods are often presented in the literature from a purely modeling point of view and they are mainly tested only on small-scale examples.

One of the main targets of this paper is to focus on the industrial applicability of existing scheduling solutions and provide some ideas or guidelines on how the gap toward industrial applicability could be reduced or even closed. As the topic itself is much too broad to be completely covered in one paper, we mainly focus on the activities done within the process systems engineering (PSE) community. In an industrial environment it is crucial to be able to connect any planning solution into the existing information systems. Often there are close interdependencies to other planning systems and control decisions and thus it is also relevant how to integrate all these together into a functional system (see Fig. 1.1).

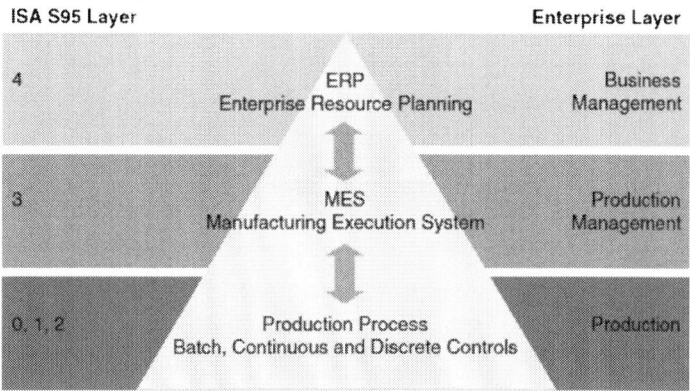

Figure 1.1: Hierarchical automation pyramid.

Production scheduling is mostly considered as a functionality of the Manufacturing Execution System (MES) in the production management layer. The various layers need to be able to interact and exchange information. For instance, production orders often come from ERP systems and a scheduling system may need to know the equipment status and condition and this information can be found in the production layer. In short, the more complex a scheduling system is, the more information must be collected and managed in order to produce feasible schedules that are also valid for production. It needs to be highlighted that the traditional automation pyramid in Fig. 1.1 may not work in all environments, while a strict hierarchical structure may provide obstacles for a true horizontal integration, e.g. with recipes under development (lacking definition).

The significance of integration and systematization of production management components such as scheduling has dramatically increased during the last years. The market for Manufacturing Execution Systems (MES) is expected to grow faster than that for base automation. The global MES-market trend can be seen in the diagram of Fig. 1.2 (Frost & Sullivan, 2010). As manufacturing sites become more flexible, complex and interconnected, the role of optimization will be more important than ever and scheduling optimization per se can be expected to be crucial in maintaining the profitability at many production plants. This also puts more pressure on improving the interfaces and plug-ability of scheduling and optimization solutions as a part of MES-systems.

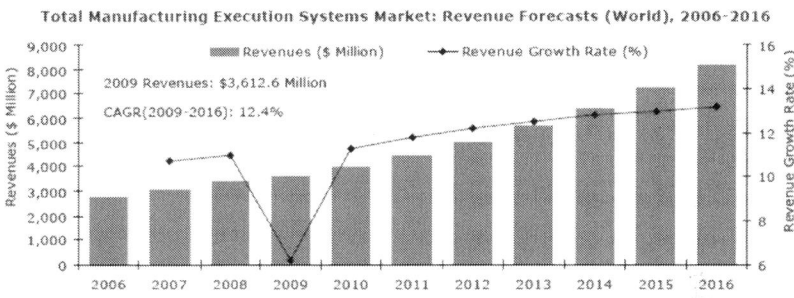

Figure 1.2: Forecasted market growth of MES-systems.

This paper first outlines the major classes of production scheduling problems, succeeded by lessons learned highlighting some industrial requirements. It briefly presents major scheduling models and solution methodologies and thereafter brings them together in the problem context, followed by a discussion on general vs. specific approaches to scheduling, where the methods are compared with each other by highlighting their major strengths and weaknesses. Special focus will be put onto relevant deployability issues – something so far neither discussed much within the scheduling literature – nor in reviews. Also the practical value and benefits of using optimization approaches for planning and scheduling will be discussed through a number of examples. Before conclusion the main emerging industrial challenges are discussed.

CLASSES OF PROBLEMS

Preliminaries

While there are many different classes of problems, the general production scheduling problem can be posed as follows. Given are:

- Production facility data; e.g. processing unit and storage vessel capacities, availability of utilities, and manpower.
- Detailed production recipes; e.g. processing times/rates, utility, and material requirements, mixing rules.
- Equipment unit suitability to carry out specific tasks.
- Production costs; e.g. raw materials, utilities, setup, cleaning.
- Production targets or orders with due dates.

The main goal is to find a schedule that satisfies all resource constraints while minimizing the total cost. Other objective functions include the minimization of makespan, earliness or lateness, as well as the maximization of throughput or profit. The major decisions are: (i) selection and sizing of tasks (batches or lots) to be carried out (batching/lot-sizing); (ii) assignment of tasks to processing units; and (iii) sequencing and/or timing of tasks on each unit (see Fig. 2.1 for an illustration). It is also important to note that scheduling can be used both off-line, to study the design of manufacturing facilities, and on-line, to plan production activities. When used off-line, finding optimal solutions is more important than computational performance, while when used online, computational efficiency is a key consideration.

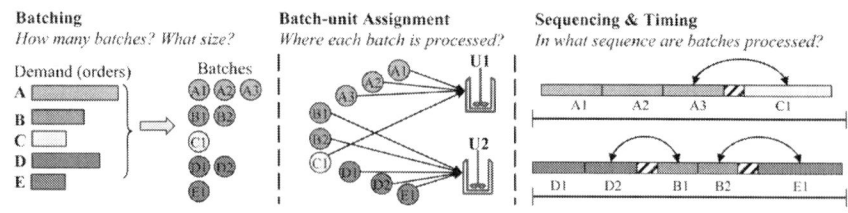

Figure 2.1: Major decisions in batch process scheduling. Continuous process scheduling involves a similar set of decisions (lot-sizing assignment sequencing).

In the process industries, scheduling appears in a wide range of areas: from the oil industry (e.g. transportation and distribution of crude oil), to the pharmaceutical and specialty chemical sectors (e.g. batch scheduling for optimal utilization of shared multiproduct facilities). Furthermore, scheduling applications arise in very different production facilities, as well as in different market and business environments. In this section, we attempt to classify scheduling problems in terms of four major attributes: (i) market environment, (ii) interaction with other planning functions, (iii) production facility, and (iv) specific processing characteristics. However, we want to stress that this classification is by no means unique. The interested reader can find details in Méndez, Cerdá, et al. (2006) and Maravelias (2012).

Market Environment

The market environment within which a company operates affects the scheduling of its plants in at least two ways. First, the volume and variability of product demand determines the regularity and frequency in which it is produced. On the one hand, production of high-volume products with relatively constant demand as well as high volume intermediates can be based on demand forecasts, rather than specific orders. As a result, the typical production goal is to maintain a certain level of stock (make-to-stock) and/or to generate a schedule that can be repeated periodically, referred to as *cyclic scheduling* or *production wheel*. On the other hand, production of chemicals with irregular and/or small demand is scheduled *as needed* (make-to-order), which typically means that a short-term schedule should be generated frequently (Pochet & Wolsey, 2006). The demand vs. capacity trade-off often determines the overarching production goal. For example, if the demand is higher than the production capacity, then the goal is to maximize the throughput or profit or minimize backlogs. If there is unused capacity, then the goal is typically to minimize the total cost, which may translate into for instance, minimization of production cost or earliness.

In summary, the market environment determines the type of problem as well as the overarching production goal. It is important to note, however, that the above considerations are not necessarily unique for a given facility. For example, it is common to have both make-to-stock and make-to-order products produced in the same plant; it is also common to have changes in the scheduling objective with demand fluctuations, contract changes, etc. Thus, different classes of problems may arise even within the same manufacturing facility.

Planning Functions

Production scheduling is only one component of supply chain management and therefore only one of the many planning functions carried out by a manufacturing enterprise (Kallrath, 2002, Maravelias and Sung, 2009, Shobrys and White, 2002 and Stadtler, 2005). As a result, other planning functions or decisions made at different planning levels affect scheduling decisions and vice versa. For example, procurement planning determines the delivery of raw materials (i.e. release dates), while distribution and demand planning determine how demand is satisfied (i.e. orders and due dates), both of which are inputs to scheduling. More importantly, production planning (medium-term) decisions determine not only inputs to scheduling (e.g. production targets), but also the type of problem. Specifically, the type of decisions made at the production planning level determines what types of decisions are left to be made at the scheduling level. For example, if both production targets and lot-sizing (or batching) decisions are made at the planning level, then scheduling involves only assignment and sequencing decisions (see discussion in Section 2.1); if only production targets are fixed, then scheduling may include batching (lot-sizing), assignment and sequencing. Furthermore, the goal in production planning (e.g. *push* vs. *pull* policies) may also determine the goal of scheduling. Finally, we note that the interaction among the various planning functions is not unique across an industrial sector, not even across plants of the same company. Rather, it depends on the culture and the information technology systems in place; e.g. enterprise resource planning systems (ERP).

Production Facility

The aspect that has received the most attention in the literature in terms of problem classification is the type of production facility. The main aspects here are: (i) the process type (batch vs. continuous); (ii) the production environment; and (iii) the type of operations (e.g. production vs. material transfer operations).

Batch vs. Continuous processing

The decisions in batch and continuous processes are for the most part the same: batching (size and number of batches) in the former corresponds to lot-sizing in the latter; while assignment (of batches/lots to units) and sequencing (between pairs of batches/lots) decisions are identical. However, the scheduling of continuous processes may involve production rate decisions. The two main differences between the two types of processes are the nature of capacity restrictions and processing times, and the manner in which these processes affect inventory levels. In batch processing, capacity is essentially a bound on the amount of processed material (kg), i.e. it affects the number and sizes of batches. Processing times can be fixed or depend on the batch size. Thus, the schedule for a batch facility involves multiple batches with specified processing times. Capacity constraints in continuous processing refer to processing rates (kg/h), and processing times are not typically constrained. This implies that capacity constraints do not affect the number of lots; i.e. a given demand can be satisfied running a single *long* lot (campaign) or multiple *short* lots or anything in between. Note that this is generally true even when restrictions on processing times are present. Hence, the scheduling of continuous processes offers additional degrees of freedom. In terms of inventory, batch processes lead to changes at specified time points or during time intervals (*filling* and *drawing*) with respect to the start of the task, while continuous processes lead to inventory depletion (accumulation) of inputs (outputs) throughout the execution of a task (Fig. 2.2). This difference results in different modeling approaches. Finally, we note that scheduling applications often involve a variety of *other types of* operations (e.g. semi-batch tasks, material transfers).

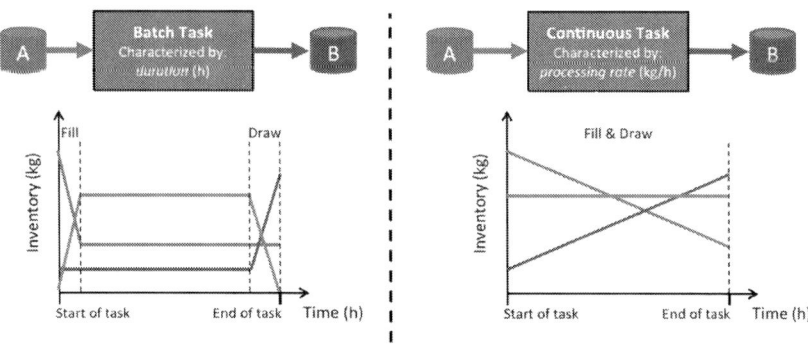

Figure 2.2: Batch vs. continuous processing.

Production Environment

The most common approach is to classify facilities as *sequential* or *network*. *Sequential* is a facility where the identity of a batch (or lot) must be preserved. In other words, a batch can neither be mixed with other batches nor split toward multiple downstream batches, which implies that the input/output of a batch should be produced/consumed by/from a single batch. *Network* is an environment where multiple batches can be mixed to form the input of another batch or the output of a batch can be consumed by multiple batches. Sequential environments are in some respects similar to discrete manufacturing facilities (e.g. car, semiconductor), where a discrete item flows through different stations of the manufacturing process. They can be further classified into the following subgroups, which, not surprisingly, are the counterparts of known scheduling problems in the operations research literature (see Fig. 2.3):

- *Single-stage* (Fig. 2.3a): Production environment with only one processing stage consisting of M parallel units; each batch has to be processed on exactly one unit.

- *Multi-stage* (Fig. 2.3b): Production environment with $K \geq 2$ processing stages; each stage $k \in \{1, \ldots, K\}$ consists of M_k parallel units; all batches have to be processed in the same sequence of stages and on exactly one unit at each stage. If $M_k = 1$ for all k, then the facility is a flow-shop. In the OR literature, the multi-stage production environment is called *flexible flowshop*.

- *Multi-purpose* (Fig. 2.3c): Environment with multiple stages where multipurpose units are suitable to carry out different operations; and different products go through a different (sub)set of stages in a product-specific sequence. In the OR literature, this facility is referred to as *flexible jobshop*, and if all stages for all products have one suitable unit, then we have a *jobshop*.

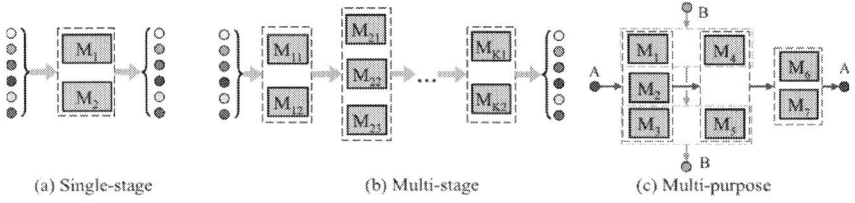

(a) Single-stage   (b) Multi-stage   (c) Multi-purpose

Figure 2.3: Types of sequential production environments with sets of products, $i \in I$, and units, $j \in J$. (a) All products have to go through a single stage; product-unit suitability can be considered using subsets J_i. (b) All products have to go through stages $k \in K$, where stage k consists of units $j \in J_k$ with $Jk \cap Jk' = \emptyset$. (c) Products have to go through product-specific stages (e.g. A and B have to go through 3 and 2 stages, respectively); $J_{i,k}$ is the subset of multi-purpose units suitable for stage k of product i (e.g. $J_{A,1} = \{M1, M2, M3\}$, $J_{B,1} = \{M1, M4\}$).

In this paper we use the term *multi-purpose* to describe a sequential facility which is a generalization of the multi-stage structure. However, this term has also been used to describe *network* environments, because the term *multi-purpose* also reflects the suitability of equipment units to perform different tasks, which means that they can be present in both *sequential* and *network* environments. We note that the same unit can be suitable for operations in multiple stages, as is typically the case in semiconductor manufacturing.

The distinction between network and sequential environments is important because it leads to different types of modeling approaches (Section 4). Most methods developed to address problems in sequential environments are based on the modeling of a batch, while methods for network environments are based on material and/or resource balances. The former typically include only assignment and sequencing decisions, while the later include batching, assignment and sequencing decisions.

Finally, we close with two remarks regarding the aforementioned classification. First, a facility does not have to be sequential or network. In fact, most facilities include processing stages where batch integrity should be preserved (e.g. manufacturing of pharmaceutical active ingredients) and stages where mixing is allowed (e.g. tablet manufacturing). Thus, the aforementioned distinction is more relevant for specific operations or processing subsystems rather than entire facilities. Second, it is the presence of material handling restrictions, rather than the structure of the facility, that determines the type of production environment (see Maravelias, 2012).

Operation Types

This paper focuses on *production* scheduling, that is, scheduling of operations that involve some type of transformation of inputs into different outputs. However, there are many operations in the process industries involving *material transfer* rather than *production*. Two important problems of this type are pipeline and crude oil scheduling (e.g. see Cafaro and Cerdá, 2004 and Mouret et al., 2009). While material transfer is typically considered independently from production (e.g. pipeline scheduling), it is important to note that production scheduling applications often involve material transfers; e.g. material charging and discharging in batch processing, use of cranes in steel manufacturing, etc.

Processing Characteristics and Constraints

Production facilities, and therefore the corresponding scheduling problems, are very different in terms of processing characteristics and constraints. In addition to the aspects introduced in the previous subsection, the scheduling of a facility may involve the consideration of features such as auxiliary equipment, utilities, labor, unit setups, changeovers, storage and transfer constraints. In this subsection we outline three of these aspects. A more thorough presentation can be found in Méndez, Cerdá, et al. (2006).

General Resource Constraints

In our discussion so far we have assumed that the major resource consideration is processing unit availability; that is why we focused on task-unit assignments and task-task sequencing on a unit. However, a facility may involve a wide variety of shared resources, ranging from auxiliary units (e.g. piping and storage vessels) and transfer equipment (e.g. cranes) to utilities (e.g. steam, water, cleaning-in-place) and labor. From a modeling point of view, resources can be discrete (e.g. labor) or continuous (e.g. steam), and materials can also be viewed as resources.

Set-Ups

In batch processing, setups are *preparatory* activities that have to be carried out before a processing activity starts. They may involve processing unit re-tooling, cleaning, and sterilization. In continuous processing, setups may correspond to transitions between two steady states (e.g. grade transition in polymer manufacturing). In both cases, they can be sequence-independent or sequence-dependent (often referred to as changeovers). They almost always imply some downtime during which the unit cannot be used or the output is off-spec, and often a cost. Setups are sometimes defined in terms of product families; i.e. setups/changeovers are necessary only between products of different families. Furthermore, setups may require shared resources such as cleaning-in-place (CIP), which means that they are resource-constrained, just as the processing tasks.

Material Storage and Transfer

The consideration of storage constraints is one of the main differences between discrete manufacturing and chemical production scheduling. There are two types of storage constraints: (i) constraints on the duration a material can be stored (in a processing unit or storage vessel), and (ii) storage capacity constraints (number and size of storage vessels). The combination of these two aspects is often referred to as storage policy (see Sundaramoorthy and Maravelias, 2008a and Sundaramoorthy and Maravelias, 2008b). The handling of storage is often treated in conjunction with material transfer activities. For example, in many facilities (e.g. metallurgical operations) the transfer of material is highly constrained because it requires shared resources such as cranes. Also, material transfer activities are often the major activities to be scheduled. Finally, resource-constrained material transfer activities often result in limited connectivity among processing units, and limitations in the number of storage vessels or units a unit can be simultaneously connected to. They may also require that material charging and drawing operations should be modeled as separate activities.

Before going through the models and methods, some practical implementation aspects are discussed in the following section.

PRACTICAL IMPLEMENTATION

Successful industrial application of scheduling technology depends on resolving many practical issues, such as change management, deployability and ease of use, application development and maintenance. To explore these issues, in the following we will recount the lessons learned from successful applications and then discuss issues around software integration and commercial tools.

Lessons Learned from Successful Industrial Implementations

This section provides a list of practical issues that must be addressed for the industrial application of advanced production scheduling technology. We will review the scheduling environment encountered in a variety of businesses, spanning continuous, batch, and discrete part manufacturing processes, as well as commodity and specialty chemicals and advanced material products. Special focus will be put on the aspects of the scheduling solutions that led to success across this diverse set of problems and the implications they have on the scheduling model or algorithm used.

Role of the Production Scheduler

The responsibilities of a production scheduler lie at the interface of the business supply chain function and manufacturing. In this position the scheduler is faced with supporting the different and sometimes conflicting goals of each organization. In some cases site logistics, responsible for incoming raw material deliveries and outgoing customer shipments, may be a third organization for the scheduler to support. Except for the simplest cases the role of production scheduler is a full time job. Generally the scheduler attempts to produce a feasible schedule that achieves the goals below in the following order:

- Operate safely
- Meet ongoing customer orders
- Meet inventory targets for replenishments for make to stock products
- Lower manufacturing, inventory and logistics costs
- Improve their work process

In the absence of an optimization tool, the production scheduler does not have an objective function to balance the financial implication of goals 2–4 and in turn try to manually satisfy them in a simultaneous fashion.

Information Available to the Production Scheduler

The data used by the production scheduler is usually spread across the business ERP system and various spreadsheets, databases and process control systems. Some of the data may be inaccurate, incomplete, or inconsistent. For instance the availability of a tank truck is not necessarily accurate, plant run rates used in ERP planning may not agree with those assumed by manufacturing, etc. In this environment the scheduler learns what data is most reliable, whom to call for supplemental information, and how to hedge their decisions by tactics like padding the schedule or carrying higher inventory.

Day-to-Day Production Scheduling

On a typical day the production scheduler may encounter several scheduling challenges originating in any of the three functions: supply chain, manufacturing and logistics. Examples of these are: rush customer orders, late deliveries or customer pickup, equipment breakdown or process delays, scheduled maintenance and product quality problems.

The frequency and impact of each of these disruptions depend on the nature of the product, process and markets served.

Generally, schedulers develop and publish a schedule to manufacturing on a regular basis (every two days, once a week, etc.) and then monitor ongoing circumstances to determine if minor adjustments to the schedule are needed or if a complete new schedule needs to be developed. The construction of a schedule can be an iterative process involving negotiations with manufacturing, supply chain or logistics and possibly also significant manual data extraction and transcribing from data sources to the scheduling tool. The whole process is very time consuming and open to data errors. The most popular tool used for manual scheduling is Microsoft Excel. Here data can be assembled from many sources, simple mass balances calculated, schedules manipulated by cut and paste actions, and a schedule formatted to the liking of manufacturing.

The Business Opportunity in Improved Scheduling

For businesses with very challenging production scheduling environments the opportunity for improvement is well recognized even if the solution is not. Businesses routinely track a variety of metrics that are directly impacted by the quality of production scheduling. These include customer service levels, sales volume, manufacturing asset utilization, manufacturing costs, inventory holding costs and rush delivery costs. If the process can be somewhat automated and less reliance placed on the scheduler's experience then scheduler turnover is less of an issue. Thus the challenge is not identifying the opportunity but accurately estimating the impact of improved scheduling, which is important for establishing alignment of all stakeholders, committing resources for implementing the change, and gaining acceptance of the change from those impacted.

Elements of Successful Implementation

The successful implementations of advanced production scheduling at Dow have been driven by both effective project execution and capable technology. Regardless of the particulars of the scheduling problem certain key success drivers seem to be common.

Quick proof-of-concept study: At the beginning of any implementation several things must be demonstrated:

- Expected value of improving the scheduling process
- Ability of an advanced scheduling approach to solve the problem
- Availability of data needed to develop the advanced solution
- How the information produced by the advanced solution can be integrated into the existing scheduling process and information system

Given the many competing priorities faced by a business considering an advanced scheduling solution it is critical that a proof-of-concept study be concluded in a short time frame, ideally two to four weeks. However, the availability of data is usually a bottleneck. The modeling approach should be flexible in describing a variety of scheduling problems and easily modified as additional information becomes available. It is also important that the modeling paradigm be easily understood by the business stakeholders. It is extremely advantageous if the solution approach can evaluate a historical schedule. This allows an improved schedule to be compared to the historical counterpart for evaluating the expected benefits of implementing advanced scheduling.

Effective System Integration: The scheduling model or algorithm is only a piece of the overall scheduling application. Production scheduling is a data intensive exercise and the success of an implementation relies on the quality of the data used by the model and the effective transfer of data to and from the model. The software infrastructure supporting the execution of the model must be capable of communicating with the business ERP system and any number of maintained spreadsheets and databases. Ultimately the model output must be made available to the scheduler in a flexible user interface. A significant effort is required to either adapt commercially available linking software or to develop such capability in house. Advanced scheduling approaches that can be leveraged across a number of implementations allow a greater return on the investment made in the creation of the software infrastructure. Likewise, subsequent implementations will be accelerated.

Effective User Control: The dynamic nature of the production scheduling environment and the difficulty in incorporating all business considerations in a scheduling model make it difficult to completely automate the generation of a schedule. It is therefore appropriate to design a system where the production scheduler can manually modify some model input data or adjust the resulting schedule to address unsolved business issues. The scheduler may also adjust the schedule as it is carried out due to minor disruptions or schedule drift. These situations require an effective user interface. The functionality of a user interface can be wide ranging and describing all the features schedulers find useful is beyond the scope of this discussion. There are several functions that lead to success:

- Interactive Gantt chart (see Fig. 3.1.) depicting the sequence of operations across each equipment unit. The schedule can be modified by rearranging the sequence of operations while important material balances and customer order fulfillment data are automatically updated.

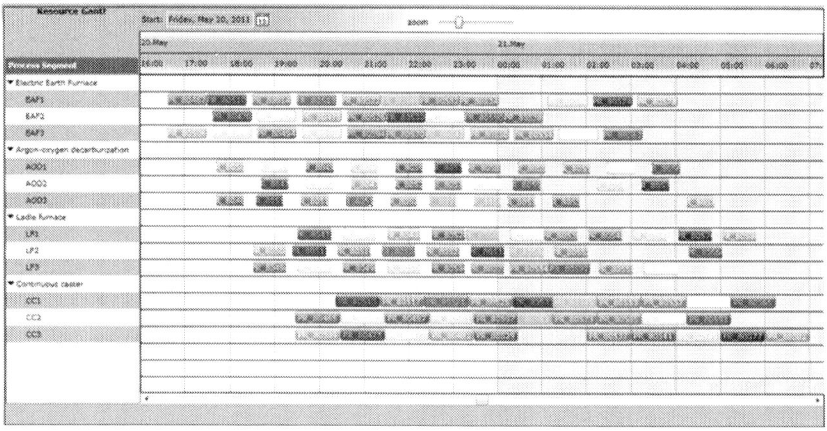

Figure 3.1: Gantt chart for presenting the production schedule.

- Trend charts displaying the expected evolution of inventories
- Easy viewing and manipulation of input data to the model
- Comprehensive reporting on order fulfillment and the scheduled operations that are associated with filling each order or replenishment requirement

These functions can be achieved if the underlying modeling approach can:

- Model individual unit operations at any desired level of detail
- Consider orders and inventory replenishments with fixed or a range of due dates

- Recognize and track the operations that are associated with particular orders
- Model inventory over time including intermediate inventory
- Fix part of the schedule and model maintenance activities and shift work.

Easy Application Development and Maintenance: The description of the scheduling problem often changes during the initial stages of implementation. The original problem formulation resulting from the proof-of-concept study is continually modified and enhanced as a more complete understanding of the problem emerges during testing and validation. Early use of the scheduling solution also tends to uncover necessary changes to account for unanticipated circumstances. In fact, development is usually accelerated if frequent iterations occur with the business stakeholders. Later, eventual changes in the production process or products require ongoing maintenance. This implementation phase works best if the production scheduler is empowered with the ability to add new products, revise cycle times and run rates or other non-structural changes to the model.

Application development and maintenance are facilitated by a modeling approach that has the following additional characteristics:

- Easy to instantiate new model elements from existing ones
- Scheduling heuristics can be represented in the model or the logic directly added to the model
- Data representing the scheduling problem can be separated from the underlying equations or algorithm of the scheduling engine (fully parameterized model)

The most interesting performance metrics to production schedulers are: the feasibility of the schedule, the quality or improvement achieved by the schedule, and the time needed to generate and report a schedule. The schedule must be feasible, i.e. taking into account all important plant constraints; otherwise it will lack credibility with the scheduler. However, it is assumed that the scheduler may fine tune the schedule produced by the model. This also provides the opportunity to balance model complexity with speed to report a solution. In the vast majority of cases the performance of an advanced scheduling tool is measured against an incumbent solution, so if the improvement results in significant financial gains for the business, the schedule does not need to be "proven optimal". Therefore solution approaches that can be terminated once an acceptable level of schedule quality is found offer a distinct advantage over those that do not. The time to report a schedule should not exceed 20 min, in some case much faster response times are required if the scheduling environment is highly dynamic or the scheduling horizon is short. Generally, solution times less than 5 min are preferred as this allows the scheduler to perform scenario testing. Decomposition algorithms have been found to be very useful in achieving desirable solution times.

The key lessons learned from various successful applications can be summarized as follows:

- Stakeholder alignment is critical. The production scheduler, his management, and the aligned groups need to all agree on the need and value of the solution (early proof-of-concept study).

- A generic modeling approach is preferred. It accelerates project timelines, notwithstanding any decomposition techniques or other methods that may ultimately be used to deliver the final solution.

- Assume the scheduler will adjust the schedule. Do not try to model all scheduling contingencies. The schedule needs to be feasible with respect to the most important scheduling constraints.

- Focus on the needs of the scheduler. In the end, the scheduler must feel that their job has been enhanced and simplified with the advanced scheduling technology, or the solution will be abandoned. The scheduling model or algorithm is only a part of the solution. Effective user control of the scheduling model is critical.

Generic Deployment Aspects

Recalling Fig. 1.1, an industrial process automation system is hierarchically organized such that the regulatory control level directly interacts with the production process equipment and devices. This may be coordinated by a supervisory control layer and more complex control and optimization related tasks are handled by real-time optimization (RTO). Planning and scheduling lies from the functional point of view between RTO and ERP and is sometimes partly or fully embedded into one of the two layers. So it is crucial that the various layers can be integrated and above all communicate.

As any solution must play a part in a more complex IT-infrastructure, implementing a scheduling optimization solution in a real industrial environment opens up new challenges, most of which have been already discussed above. The most important requirements that apply to any kind of scheduling systems that should be functionally integrated to an industrial software-environment are:

- The scheduling solution must be robust, scalable and stable for any case, meaning that there are fallback strategies. In cases where the optimization fails, at least a feasible schedule must be reported.

- In running production, every schedule is, in fact, a re-schedule that builds upon a previous schedule. In case of sudden disturbances, a new adjusted solution should be generated efficiently taking into account the current state of the process.

- The solution must be maintainable, both in terms of updating the model and in keeping up with major software updates in the plant.

- It should be possible to easily integrate the solution and link it to other applications, for instance ERP-software or other tools that provide the overall equipment efficiency or calculate and predict the total production costs.

- The cost to implement solution should ensure a good return-on-investment (ROI), which is also affected by the related software license and development costs

- Few generic guidelines about the IT infrastructure around production scheduling can be found in the ISA-95 standard (ANSI/ISA, 2005) and in ZVEI (2011). For more information about Collaborative Process Automation System, its components and integration of these, we refer the reader to Hollender (2009) andEngell and Harjunkoski (2012).

BASIC TYPES OF SCHEDULING MODELS

Given the wide variety of aspects and features that may be encountered at the process industries, it is not surprising that a large number of different scheduling models can be found in the literature. In this section, we provide an overview of mathematical programming formulations by describing and illustrating the key variables and constraints, directing the reader to the appropriate references for further details. Similar types of models can be generated using other modeling and solution paradigms. All models presented in this section are mixed-integer linear programming (MILP) models, unless blending, time-dependent inventory costs or periodic modes of operation need to be considered, which turns the model into an MINLP (Section 4.6). In Section 4.1, we present a high-level classification of models, which is followed in Section4.2 by the different problem representation approaches. The critical issue of modeling of time is discussed in Section 4.3, followed by representative mixed-integer programming (MIP) formulations in Sections4.4 and 4.5.

Model Classification

Scheduling models can be classified in terms of: (i) optimization decisions, (ii) major modeling elements, and (iii) the modeling of time. In terms of decisions, most models for sequential facilities include assignment of tasks to processing units, sequencing between pairs of tasks assigned to the same unit, and/or timing of tasks (Maravelias, 2012). Models for network facilities include batching, task-unit assignment (or task-resource allocation), and timing decisions.

In terms of modeling elements, in addition to production units, models are typically expressed in terms of*batches* or *material amounts*. Problems in sequential production environments, where the same batch is assumed to be processed in different stages, have been traditionally modeled using batches. On the other hand, problems in network-type environments, where different materials as well as batches of the same material are mixed and a batch can be split, have been modeled using materials (represented either as states or resources). This alternative will be further discussed in the problem representation section.

The modeling of time involves: (a) the selection between precedence-based and time-grid-based approaches; (b) the specifics of the precedence-based (e.g. local vs. global precedence) or time-grid-based (e.g. common vs. unit-specific grid) approach; and (c) selection between continuous- and discrete-time representation. In general, scheduling models for sequential environments employ a batch-based approach combined with precedence-based or time-grid-based modeling of time (Section 4.4), while models for network environments employ a material-based approach combined with time-grid-based modeling of time (Section 4.5).

Problem Representation

Modeling a scheduling problem demands a problem representation, which can be viewed as an abstraction layer between the real problem entities (e.g. orders, units, and stages) and the model entities (variables and constraints). For instance, chemical manufacturing can be quite complex; a simple *unit procedure* may consist of multiple *operations*, which in turn consist of multiple *phases*. However, scheduling models can be developed based on a simplified representation of the manufacturing process. There are two broad classes of such representations.

The first, which is used to represent problems in sequential production environments, is based on (i) a set of distinct production *stages*, (ii) units in each stage, (iii) batches or products and orders (Fig. 2.3). This representation can also be used when there is material mixing or splitting that can be ignored (e.g. removal of waste material). If the batching problem is solved before scheduling, then the scheduling problem involves a given set of batches, which have to be processed in different stages. Problem data in this case include batch routings (i.e. sequence of stages), batch-unit suitability, processing times of batches, and batch release/due dates. Batches may have different routings, a unit may belong to multiple stages, and units may be suitable for the processing of only a subset of batches. Also, since the batch size is fixed, processing times are constant. If the batching problem is not solved a priori, then the representation is based on product orders (instead of batches), stages and units. Problem data include product routings, product-unit suitability, processing times of products (which can be functions of batch size), and orders (amounts and release/due dates). The vast majority of models developed to address problems in sequential facilities are based on this representation.

The second type of representation is more generic and relies on the modeling of materials, tasks, units and utilities. The state-task network, STN (Kondili, Pantelides, & Sargent, 1993) and the resource task network, RTN (Pantelides, 1994) are the two most popular representations of this type. The STN represents processes as a collection of different material states that are transformed by tasks (Fig. 4.1) carried out in units. Information about the processing equipment required to perform the tasks is specified as a set. In STN, there can be three separate entities besides tasks: states, units and utilities. The RTN is a unified representation where these entities are treated as resources produced and consumed by tasks (Fig. 4.1). Both STN and RTN can be extended to represent storage vessels and alternative material locations, as well as different equipment states (e.g. clean, dirty, ready to process). Both STN and RTN representations were originally used for problems in network environments but have recently been used to address problems also in other environments (Sundaramoorthy and Maravelias, 2011b and Velez and Maravelias, 2013d).

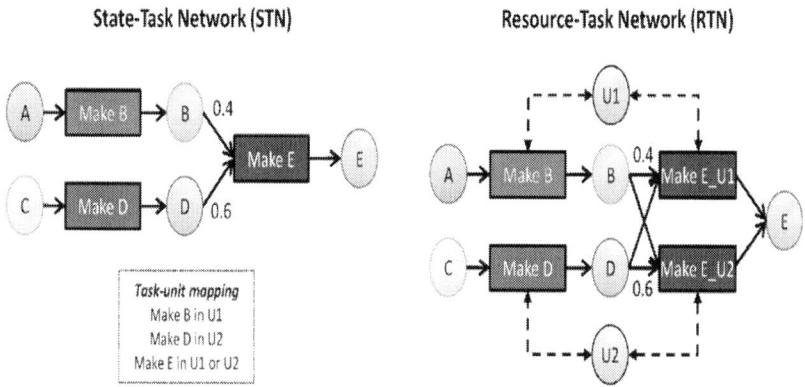

Figure 4.1: Problem representations for scheduling in network production environments.

It is important to note that a problem representation can be used as a basis for the development of different models and solution approaches. For example, the representation based on stages, units, and batches has been used to develop a wide range of MIP and CP models as well as scheduling algorithms. Similarly, RTN and STN representations have been used to develop different MIP, CP and TA-based models (Fig. 4.2).

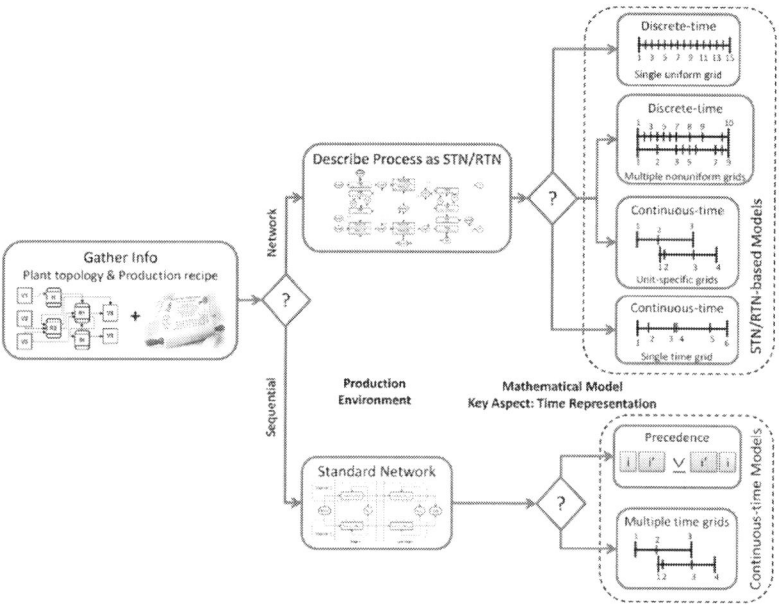

Figure 4.2: Proposed strategy to address scheduling problems.

Modeling of Time

Precedence-based models, mainly used for the scheduling of sequential environments, include unit-batch allocation and batch-batch sequencing constraints (Pinto & Grossmann, 1998). The former is modeled via assignment binary (or logic) variables ($Yi\,m$ = 1 if batch i is assigned to unit m) and constraints that ensure that each batch is assigned to exactly one unit at each stage. Sequencing constraints, ensuring that each unit processes simultaneously at most one batch, are modeled using precedence binary variables. There are two types of precedence variables for a pair of tasks, i and i ', processed on the same unit: immediate, where $X_{i,i'}$=1 if batch i' immediately follows batch i ; and (ii) global, i.e. $X_{i,i'}$=1 if batch i' is processed after (but not necessarily immediately after) batch i in a unit (Fig. 4.3A). In both cases, the number of precedence binaries and constraints increases quadratically with the number of batches since both variables and sequencing constraints are expressed for pairs of batches. In general, global precedence models require fewer binary variables and are thus better options than their immediate precedence counterparts, but on the downside it is cumbersome to identify subsequent tasks, making it considerably harder to account for sequence-dependent costs and to enforce or forbid certain processing sequences.Kopanos, Laínez, and Puigjaner (2009) and Kopanos, Méndez, and Puigjaner (2010) have overcome this issue efficiently, by combining unit-specific immediate and general precedence variables in the same model.

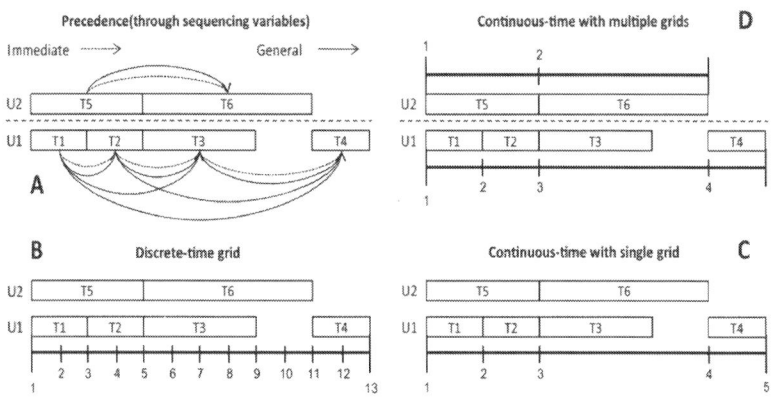

Figure 4.3: Different concepts for representing the time in scheduling formulations.

We use the term time-grid-based to describe all modeling approaches that employ time slots/periods/points/events. Time-grid-based models rely on a mapping of the tasks onto one or more time reference grids and a wide variety of approaches can be encountered in the literature. They can be classified as discrete or continuous and can be applied to network as well as sequential environments. In discrete-time models, the scheduling horizon is divided into a finite number of typically uniform time periods of a given duration. The time corresponding to each point is thus defined a priori (Fig. 4.3B). In continuous-time models the horizon is partitioned into a fixed number of periods whose length is determined by the optimization. Continuous time-grid-based models can be subdivided into approaches that employ a single, common grid across all units (Fig. 4.3C), and approaches that employ multiple, unit-specific grids (Fig. 4.3D). A discrete-time model employing unit-, state-, and task-specific grids was also recently proposed (Velez & Maravelias, 2013a).

Since almost all variables and constraints in time-grid-based models are expressed for every time point or period, discrete-time models lead to large-scale MIP models, which however are quite tight. We should also note that the challenge in discrete time models is selecting the number of intervals so as to accommodate values of arbitrary processing times as this involves rounding of these values which in turn may introduce some error in the timing of events in the schedule. Continuous-time models lead to substantially smaller formulations since fewer periods are required to represent the schedule. Nevertheless, from a computational point of view they do not necessarily perform better than their discrete-time counterparts. Single grid models are more general, while multiple grid models become computationally more efficient if a different number of slots is used in each grid. While this may not be practical in full-space models where finding the optimal common number of events is already non-trivial, it is doable in decomposition methods (Castro, Harjunkoski, & Grossmann, 2011). Finally, there are models that combine multiple time grids or different ways to model time in order to handle constraints on transportation devices (Bhushan and Karimi, 2003 and Castro et al., 2012) as well as to account for inventory constraints in integrated planning-scheduling models (Chen et al., 2008, Erdirik-Dogan and Grossmann, 2006, Kopanos et al., 2011 and Liu et al., 2008).

Models for Sequential Production Environments

Scheduling problems in sequential environments with the features shown in Fig. 2.3 can either be tackled with precedence based or time-grid-based models. In this section, some modeling approaches are illustrated graphically. For more details we refer to the earlier mentioned review papers or to specific contributions.

Precedence-Based Models

Assume that the scheduling problem involves solely allocation, sequencing and timing decisions. Let Boolean variables $X_{i,i'}$ indicate that production order i precedes i' and $Yim = True$ when order i is allocated to unit m. If both i and i' are allocated to the same unit m, then it is either i before i' OR the other way around, see Fig. 4.4. Note that it is not strictly necessary to start i' after the end of i plus changeover and setup times since there may be other orders in between. These two constraints can be modeled as a Generalized Disjunctive Program (GDP) and easily converted to an MILP following a big-M reformulation and transformation of the logic expressions into linear inequalities (see Castro & Grossmann, 2012, for further details). Fig. 4.4 shows the generalization for any sequential facility structure where the Boolean variables have been converted into binary variables. Stage indices k/k' have been added to all variables and parameters since it is possible in a multipurpose configuration for an order to go through a particular unit more than once. The additional constraints ensure that the starting time of order i in stage $k + 1$ is greater than its ending time in stage k and that all orders are processed in exactly one unit of their specified stages.

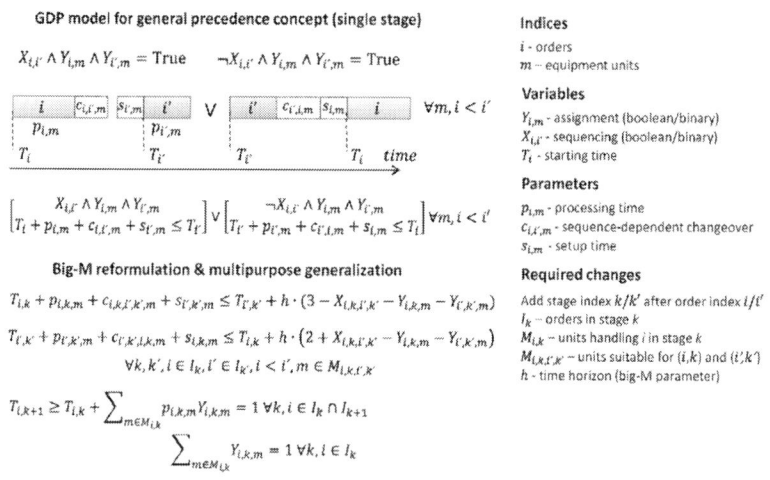

Figure 4.4: Key constraints of general precedence model for sequential multistage and multipurpose plants.

The major strength of a precedence-based model is that it can provide high quality solutions with limited computational resources, even though it may be difficult to prove optimality. Precedence models are particularly effective when partial or total pre-ordering can be performed using information such as due dates and product families, since this can significantly reduce the number of binary variables.

Release and due date information can further be used to fix variables and generate tightening constraints.

Early precedence-based models considered problems with fixed batches, unlimited storage, and no general resource constraints (Harjunkoski and Grossmann, 2002, Méndez and Cerdá, 2002 and Méndez et al., 2001). Storage constraints with fixed batches were studied by Méndez and Cerda (2003a), while a special case of utility constraints with fixed batches were discussed in Méndez et al. (2001). Simultaneous batching and scheduling in single-stage facilities was studied by Castro, Erdirik-Dogan, and Grossmann (2008) and in multi-stage facilities by Prasad and Maravelias (2008) and Sundaramoorthy and Maravelias (2008a). Sundaramoorthy and Maravelias (2008b) proposed a model for the simultaneous batching and scheduling under different storage policies. Ferrer-Nadal, Capón-Garcia, Méndez, and Puigjaner (2008)studied transfer of materials in plants with reverse flows (Fig. 2.3c) in multipurpose facilities. The same group has also tackled network structures with continuous rather than batch processes featuring tasks with variable processing rates (Capón-Garcia, Ferrer-Nadal, Graells, & Puigjaner, 2009) under the assumption of single production campaigns for each product.

Time-Grid-Based Models

Several different models can be found in the literature that can be distinguished based on: (i) specific aspects of process configuration (e.g. number of stages, parallel units, intermediate storage/waiting policies); (ii) specific aspects of time representation (e.g. unit- or stage-based time slots); (iii) number of indices in the binary variables; (iv) type of constraints associating the timing variables to slots of units of consecutive processing stages.

We now briefly describe the single-stage version of the model proposed by Castro and Grossmann (2005), which can be derived from a GDP model followed by a compact convex hull reformulation (Castro & Grossmann, 2012), see Fig. 4.5. Note that it can be assumed without loss of generality that every order requires a single slot, leading to the requirement of fewer slots to represent the solution and hence a significantly better computational performance.

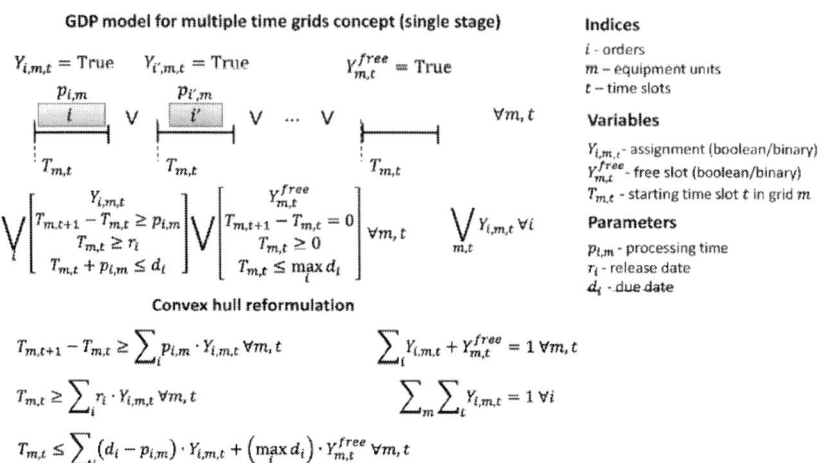

Figure 4.5: Key constraints of time grid based model for sequential multistage plants.

Assume a sufficient number of slots is specified and let $Y_{i,m,t}$ indicate if order i is assigned to time slot t of unit m. Clearly, a slot can be occupied by any order but there is also the possibility of unassigned slots (variables $Y_{m,t}^{free}$). The constraints inside the first disjunction state that the duration of any active slot must be greater than or equal to the processing time of the order assigned to that slot. The slot starting time ($T_{m,t}$) must be greater than the order's release date and the slot must end before the order's due date. The second disjunction relaxes these three constraints if the slot is idle. We also need to ensure that every order is assigned to a slot of a unit (see disjunction on the right). Conversion to the MILP format results in five sets of constraints.

The overwhelming majority of time-grid-based models for sequential environments adopt continuous modeling of time. Only the model of Sundaramoorthy, Maravelias, and Prasad (2009), which considers the simultaneous batching and scheduling in sequential environments under general storage and resource constraints, employs discrete-time modeling.

Time-grid-based models for sequential facilities tend to be tighter and computationally superior to precedence-based models despite generating larger models that are less intuitive (Castro and Grossmann, 2005, Liu and Karimi, 2007, Liu and Karimi, 2008, Pinto and Grossmann,

1995, Stafford et al., 2005 and Tseng et al., 2004). The extensive comparative studies performed in the above references and also in Castro, Grossmann, and Novais (2006) and Mouret, Grossmann, and Pestiaux (2011) clearly conclude that there is no single best performer and that even a slight change in process configuration or objective function can affect the model performance drastically. Nevertheless, one can say that it is typically better to rely on unit-specific rather than stage-specific slots featuring binary variables with three indices (order, unit, slot).

Models for Network Production Environments

The network environment represents the most complex arrangement, with tasks that can produce and consume multiple materials, batch mixing and splitting, and materials being consumed and produced by different tasks executed in different units. Thus, to accurately monitor inventory as well as utility consumption levels, a systematic method to map events that lead to changes in the state of the system (e.g. inventory levels, unit status, etc.) onto time is needed, which is accomplished using approaches that rely on time grids. In the following, we employ the RTN process representation since the unified treatment of resources allows us to focus on fewer sets of constraints. Differentiation of the resources and accounting for their interactions with tasks is achieved through structural parameters that connect the process model to the time grid based mathematical model.

Linking the Process Model to the Mathematical Model

The primary goal of the RTN process model is to identify the resources and tasks that are needed to describe the problem at hand together with their interaction. Let us assume for illustrative purposes that an instance of a task i is confined to a single slot t (Fig. 4.6). Tasks are characterized by: (i) two sets of variables, binary assignment variables $N_{i,t}$ and non-negative continuous variables $_{i,t'}$ which give the amount of material handled by the task; (ii) five sets of structural parameters. Assume for now that discrete interactions occur either at the start or end of the task, while a continuous interaction takes place throughout the execution of the task. Parameters $\mu_{r,i}$ and $\overline{\mu}_{r,i}$ are used for the discrete interactions and whenever the amount of resource r consumed/produced is independent of the amount handled by the task, as it is the case for equipment units (e.g. unit M). For material resources (e.g. A), one normally relies on parameters $_{r,i}$ and $\overline{V}_{r,i}$, linked to the continuous extent variables $_{i,t}$. Finally, parameters $_{r,i}$ hold the continuous interactions (e.g. task TC with C).

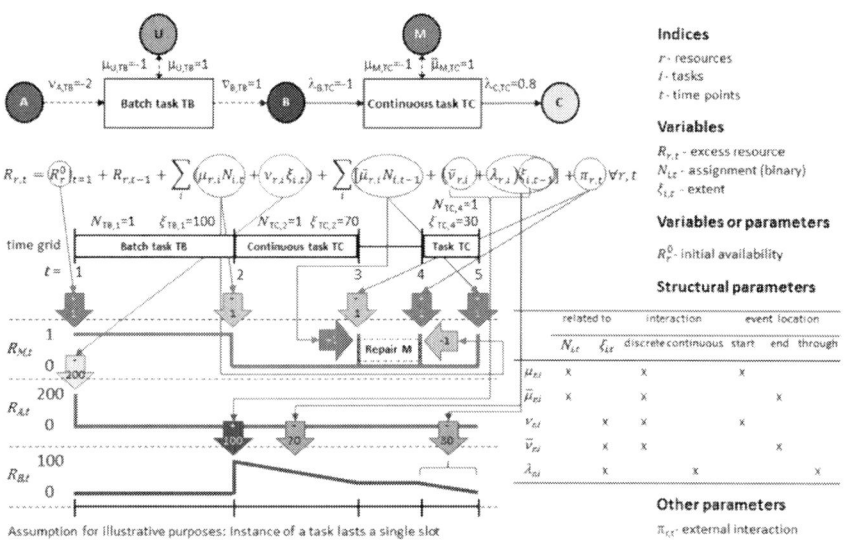

Figure 4.6: RTN entities and how the excess resource balances work.

In the RTN-based mathematical model, structural parameters come into action in the excess resource balances, see Fig. 4.6. These are multiperiod balance equations featuring excess resource variables Rrt(continuous, non-negative). Their values are typically linked to equipment availability and storage requirements for processing materials but are somewhat counterintuitive. In case of equipment units, $Rrt = 0$ means that resource r is temporarily unavailable at event point t (start of slot t, where something may start/finish). This may be due to r being already assigned to a task (e.g. TC at $t = 2$) or being under maintenance (e.g. during slot $t = 3$). The latter case is modeled through parameters ϖrt, which account for the interactions with the surrounding environment that occur during the time horizon.

To initialize the system, it is better to use R_r^0, which can either be considered as a parameter or, if one wants to evaluate if the resource is needed, as a variable. For material resources, $R_{r,t} = 0$ results in a non-intermediate storage policy. Excess resource variables must lie within given upper and lower bounds $R_r^{min} \leq R_{r,t} \leq R_r^{max}$.

In the following three conceptually different models, the RTN entities are essentially the same even though they may feature more or less indices and terms, related to the model assumptions.

Discrete-Time Models

The RTN discrete-time formulation (Pantelides, 1994) assumes fixed processing times, multiples of the chosen duration ( ) of the slots of the uniform grid. Knowing the exact location of every time point in the grid is perhaps its most important feature leading to the straightforward modeling of intermediate events such as release/due dates, equipment unavailability, time-dependent utility pricing and availability, and linear modeling of holding and backlog costs (Sundaramoorthy & Maravelias, 2011a). It thus allows for easy integration with a higher level planning model (Maravelias & Sung, 2009). Another advantage of the discrete-time formulation is that it can handle events that occur at a time   (in number of slots) relative to the start of the task and not just at the start (= 0) or end of tasks (= i). This is particularly useful to model setups and avoids the need to define subtasks, as can be seen in the example in Fig. 4.7, where utility requirements vary throughout the duration of the reaction task (e.g. steam required during the first hour and cooling water during the last three, at different rates).

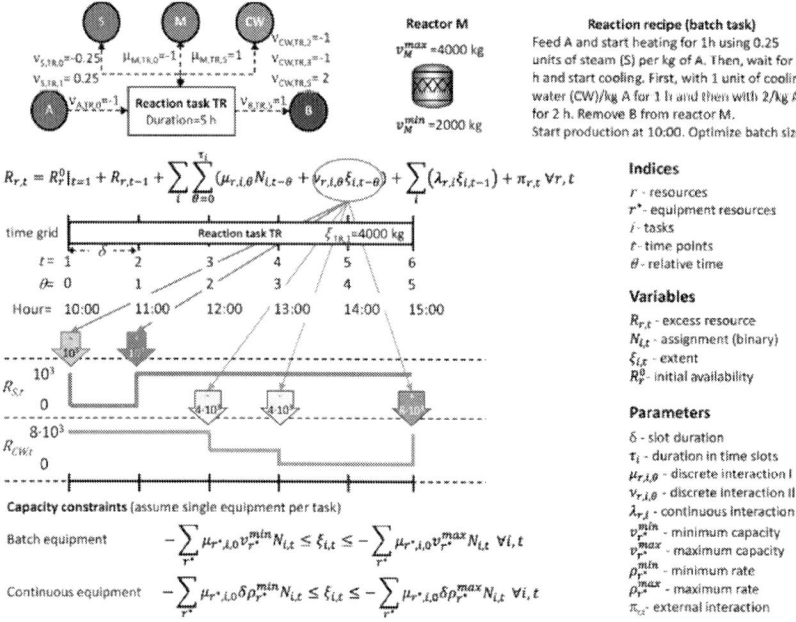

Figure 4.7: Discrete-time RTN model can handle events beyond start and end of tasks.

The major drawback is related to data accuracy. An exact model is only obtained with   equal to the greatest common factor of all *i*, which can lead to a large number of time slots. Normally, an approximate model is considered where the processing times are rounded up to a multiple of , *i* = ⌈ / ⌉, where *i* is the duration in real time units. Increasing   leads to fewer time slots and the generation of smaller problems. Other features that can lead to large discrete-time models are: (i) sequence-dependent changeovers or non-instantaneous transfer times between units of a different order of magnitude than processing times; and (ii) variable processing times either as a (linear) function of the batch size or as any value above a minimum processing time, effectively set by the availability of a bottleneck resource.

Overall, the discrete-time formulation is perhaps the most general approach, being capable of handling problems of industrial relevance. Examples include: scheduling of biochemical processes (Samsatli & Shah, 1996) and fast-moving consumer goods plants (Yee & Shah, 1997); scheduling of heat exchanger cleaning (Georgiadis, Papageorgiou,

& Macchietto, 2000); recipe optimization of a blending process (Glismann and Gruhn, 2001 and Kolodziej et al., 2013); handling steam availability in the cooking process of a pulp plant (Castro, Barbosa-Póvoa, & Matos, 2003); discrete-manufacturing scheduling (Georgiadis et al., 2005); grinding process of ceramic tile (Duarte, Santos, & Mariano, 2009) and cement plants (Castro, Harjunkoski, & Grossmann, 2009); and (iv) a liquid waste treatment network (Wassick, 2009).

Continuous-Time Models Relying on a Single Time Grid

Continuous-time models have the advantage of being accurate and more sensitive to small changes in task durations, and can thus be considered more appropriate for the integration with the lower-level control layer. The major drawback is that they are significantly more complex than their discrete-time counterpart, particularly when batch tasks are considered. The following RTN single-time grid continuous-time formulation is taken from Castro, Barbosa-Póvoa, and Matos (2001), Castro, Barbosa-Póvoa, Matos, and Novais (2004) and Castro, Westerlund, and Forssell (2009), an upgrade of Schilling and Pantelides (1996). It can handle a wide variety of features, the most important being the ability to consider both batch and continuous tasks. The STN formulation of Maravelias and Grossmann (2003a) and the model bySundaramoorthy and Karimi (2005) are other well-known continuous-time models relying on a single time grid.

The time horizon is also here divided into a fixed number of slots but contrary to its discrete-time counterpart, the exact location of the event points is determined through variables T_t. This makes it computationally harder to model due date constraints, as binary variables $Y_{t,d}$ are needed to relate the specified timing hd of due date d with the location of exactly one event point. The due date timing constraint (Maravelias & Grossmann, 2003b) written in GDP format is shown in Fig. 4.8.

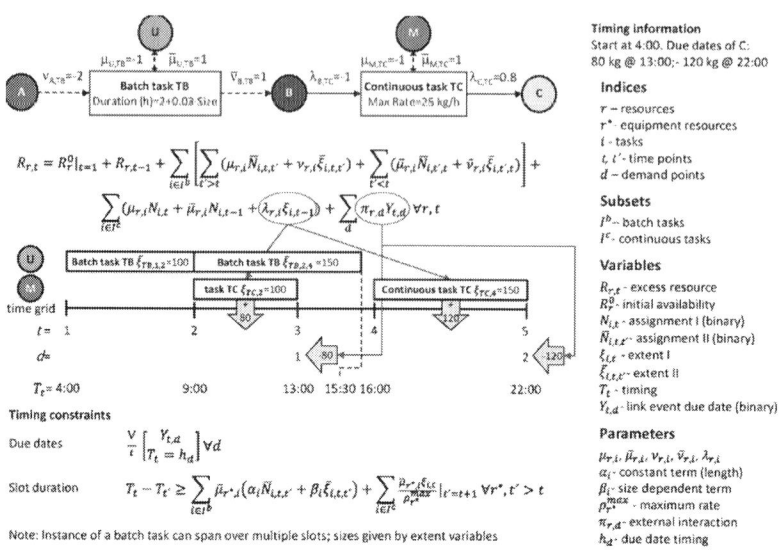

Figure 4.8: Single grid continuous-time RTN model.

The binary and extent variables associated to batch tasks now need to be characterized by two (starting and ending) times. It is assumed, without loss of generality, that instances of continuous tasks span over a single slot (Fig. 4.8). The model can easily handle tasks of variable duration being normally assumed that the duration of a batch task is given by a constant plus a term proportional to the amount of material being handled, $\alpha_i + \beta_i \bar{\xi}_{i,t,r'}$. In turn, the length of a continuous task must be higher than the task extent divided by the maximum processing rate. If the equality does not hold, we are implicitly assuming that the material is held in the corresponding unit longer than it should (e.g. batch task ending at 15:30 when the location of event point 4 is at 16:00) or that the continuous task can be processed below its maximum processing rate.

Overall, significantly fewer slots are required to represent a schedule on a continuous time grid but the amount of slots also has a stronger influence on both solution quality and the computational effort. The problem is that we do not know a priori how many slots are required to find the global optimal solution. To overcome this issue, a standard iterative procedure is used, where one keeps increasing the number of slots and solving the optimization problem as long as improvements in the objective function are observed (Méndez, Cerdá, et al., 2006). However, this must be classified as a heuristic approach since additional increments in the number of slots may lead to further improvements (Liu & Karimi, 2007). Numerous extensions of continuous-time models have been proposed to handle a range of complex processing characteristics and constraints (Castro, 2010, Castro et al., 2009a, Gimenez et al., 2009a and Gimenez et al., 2009b).

Continuous-Time with Multiple Time Grids

The performance of the single grid model is severely compromised by the need for batch tasks to spread across a large number of consecutive periods. Assigning tasks to multiple time grids lowers the required number of slots and the problem size, which may result in a major decrease in computational effort. Multiple time grid approaches are more commonly known as unit-specific.

Ierapetritou and Floudas, 1998a and Ierapetritou and Floudas, 1998b developed the first unit-specific model for network representations. Several improvements have appeared in the literature ever since. Next, we present an RTN version of this model from Shaik and Floudas, 2008 and Shaik and Floudas, 2009which allows tasks spanning across multiple slots. While called a unit-specific approach, the multiple time grids are actually associated to tasks through timing variables $T_{i,t}^s$ and

$T_{i,t}^f$. These give the start/finishing times of tasks starting/finishing at slot t and it is possible for tasks to start and end at the same slot, as can be seen for task i' in Fig. 4.9. The key timing constraints involve sequencing: (i) tasks executed in the same equipment unit (e.g. U, top-left); (ii) any pair of tasks producing and consuming a given material resource (e.g. B, bottom-left). Concerning tasks on the same unit, the GDP constraint on the top-middle ensures that if a task i starts at t and ends at $t' \geq t$, then the finishing time must be equal to the starting time plus the processing time. To ensure no overlap of tasks, the starting time of slot $t + 1$ must be greater than or equal to the finishing time of slot t. To improve the tightness, the third constraint ensures that the total duration of tasks is lower than the time horizon. As for tasks involving the same material resource, moving to the next production stage is always associated to an increase in the slot index. More specifically, if task i' finishes at slot t then the following resource consuming task i

must start at slot $t + 1$ or later. If not, the second term on the right-hand side relaxes the constraint.

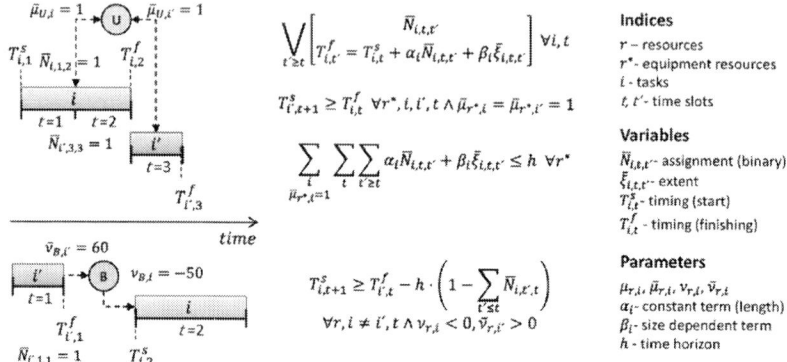

Figure 4.9: Key timing constraints of unit specific model.

The interested reader is referred to: (i) Li and Floudas (2010), addressing the major issue of specifying the optimal number of event points; (ii) Susarla, Li, and Karimi (2010) and Seid and Majozi (2012), for alternative models that are more rigorous at managing shared resources. Unit specific models have been employed to tackle large-scale industrial problems (Janak et al., 2006a, Janak et al., 2006b and Shaik et al., 2009), further details are given in Section 7.7.

Nonlinear Scheduling Models

Many production scheduling problems reported in the literature correspond to MILP models (e.g. seeFloudas and Lin, 2004 and Méndez et al., 2006a) since they assume simplified representation of the process models. For instance, fixed time horizons, fixed processing times or fixed processing rates are typically assumed. Furthermore, no additional process variables such as flows, compositions and temperatures are normally considered. However, in a number of applications such as crude oil scheduling (Lee et al., 1996 and Mouret et al., 2009), production scheduling for gasoline blending (Méndez, Grossmann, Harjunkoski, and Kabore, 2006), scheduling of ethylene furnaces (Jain & Grossmann, 1998), scheduling and control of polymerization reactors (Flores-Tlacuahuac and Grossmann, 2006, Mishra et al.,

2005, Nystrom et al., 2005, Prata et al., 2007 and Terrazas-Moreno et al., 2007), it is necessary to incorporate a description of the physical process that takes place, including differential-algebraic equations in the latter case. These nonlinear models can in fact apply to any of the basic scheduling models discussed in Sections 4.2 and 4.3.

The simplest case that introduces nonlinearity in a scheduling model, and is not related to a process model, is when cyclic schedules with infinite horizon are considered (e.g. Castro et al., 2003, Pinto and Grossmann, 1994, Sahinidis and Grossmann, 1991a and Shah et al., 1993). In this case the models are still basic MILP models like the ones discussed in Section 4.4, but the objective function is divided by the length of the cycle time, Tc, which renders such a function to be linear fractional, which is pseudoconvex. Hence, in this case the relaxed MINLP exhibits a unique optimal solution. You, Castro, and Grossmann (2009) have shown that a specialized algorithm, Dinkelbach's algorithm, can be applied effectively to solve these problems through a sequence of MILP problems (see also Pochet & Warichet, 2008).

The other type of nonlinearity that can be introduced and is not related to a process model are cyclic scheduling models with average inventory equations (e.g. Pinto & Grossmann, 1994), which have the general form,

$$AveInv = \frac{1}{2}\sum_i\sum_k Cinvf_i \left(\gamma p_{iM} - \frac{Wp_{ikM}}{Tc} \right) Tpp_{ikM}$$

where $Cinvf$ is the inventory cost, $_{piM}$ is the production rate, while the variables are Tc cycle time, Wp_{ikM} the production of product i in slot k last stage M, and Tpp_{ikM} the corresponding processing time. Note that this term is in fact nonconvex and may require global optimization methods for its rigorous solution (e.g. Horst and Tuy, 1996 and Sahinidis and Grossmann, 1991a).

The simplest, and perhaps most prevalent form of a process model is the blending equations that lead to bilinear terms as shown in the following equation:

$$\sum_{i \in I_n} (F_{i,n} P_{i,n}^j) - F_{o,n} P_{o,n}^j = 0 \quad \forall n \in N, \quad \forall j \in J$$

$$\sum_{i \in I_n} F_{i,n} - F_{o,n} = 0 \quad \forall \ n \in N$$

where F represents total flows, P properties such as concentrations or physical attributes (e.g. ROM), N is the set of nodes with blending operations, I_n is the set of stream entering node n, and J is the set of properties.

A major difficulty that arises with the above equations is that they are nonconvex, and therefore commonly give rise to multiple local solutions. Therefore, the corresponding scheduling problems become nonconvex MINLP problems that must be solved to global optimality for obtaining rigorous global optimization solutions (see Kolodziej et al., 2013 and Misener and Floudas, 2010). At the other extreme of complexity, another form of nonlinearity that can arise in scheduling problems is when dynamic process equations are added to the scheduling model which gives rise to mixed-integer dynamic optimization (MIDO) problems, This is for instance required when modeling transitions from one grade to another in the manufacturing of polymers (Flores-Tlacuahuac and Grossmann, 2006 and Nystrom et al., 2005). This is also required to model the dynamic performance of batch units for which the processing time may be a major decision variable to be determined as it defines the conversion of a given chemical reaction. In this case the most common approach is to discretize the system of differential equations through orthogonal collocation and reformulate the MIDO problem into a large-scale MINLP model, which is often very difficult to solve.

SOLUTION METHODS

In this section the most common solution methods are discussed. More information can be found in the supplementary material.

MILP Methods

As mentioned in the previous sections, alternative MILP-based formulations have been developed to address major features found in short-term scheduling problems. Those MILP models can be solved using primarily linear programming (LP)-based branch & bound (B&B) solvers, which have the advantage of providing rigorous lower and upper bounds on the solution, which in turn gives information regarding the optimality of the solution. MILP problems may be represented by the following general formulation:

$$\min z = cx + fy$$
$$s.t.\, Ax + By \geq b \quad (MILP)$$
$$x \in \Re_{+}^{n}, y \in \{0, 1\}^{p}$$

where c, f, b are vector of constants, A and B are matrices of constants, and general integer variables are represented using binary variables. B&B algorithms rely on the solution of a linear relaxation (where binary variables are treated as continuous). Broadly speaking, the algorithm starts with the solution of the initial relaxed LP problem

at the root node; an integer variable, yi, with fractional value, y_i^0, at the LP solution is chosen for branching; two additional subproblems

are generated by adding the inequality $y_i \leq \lfloor y_i^0 \rfloor = 0$ and $y_i \geq \lceil y_i^0 \rceil = 1$ to the LP problem at the root node; the LP subproblems are solved, and the procedure is repeated. The best LP solution provides a lower bound while the best integer solution provides an upper bound to the optimal solution, so the algorithm terminates when the gap between the bounds is closed. During the search, the upper and lower bounds are used to prune branches of the tree (Land and Doig, 1960 and Nemhauser and Wolsey, 1988).

While real world scheduling problems tend to lead to large-scale MILP models, truly remarkable progress has been made in recent years in the ability to solve MILP models mainly due to: (a) advances in computational resources in terms of CPU speed and memory, and (b) developments of new and improved algorithms. For example, an instance of the classical STN model (with 72 binary and 179 continuous variables, and 250 constraints), was solved in 908 s and 1466 nodes on a SUN Sparc in 1993 (Shah et al., 1993), whereas the same instance was solved in only 0.45 s using CPLEX 7.5 on a laptop IBM T-40 in 2003 (Méndez, Cerdá, et al., 2006). Algorithmic advances include (a) effective preprocessing methods, (b) use of heuristics within the B&B search, (c) generation of cutting planes at preprocessing and during the search, (d) integration of local search methods within B&B algorithms, and (e) methods that harness parallel computing resources (Atamturk and Savelsbergh, 2005, Bixby and Rothberg, 2007, Cornuejols, 2008, Johnson et al., 2000 and Wolsey, 1998). In addition, decomposition strategies and other solution methods have helped to facilitate the solution of larger MILP models. Specific solution methods for chemical production scheduling MILP models include decomposition and iterative methods (Bassett et al., 1996, Calfa et al., 2013, Ferris et al., 2009, Harjunkoski and Grossmann, 2002, Kelly and Zyngier, 2008,Kopanos et al., 2010, Papageorgiou and Pantelides, 1996 and Wu and Ierapetritou, 2003), reformulations (Janak and Floudas, 2008, Sahinidis and Grossmann, 1991b, Velez and Maravelias, 2013b and Yee and Shah, 1998), tightening methods (Burkard and Hatzl, 2005 and Velez et al., 2013), and parallel computing algorithms (Subrahmanyam et al., 1996 and Velez and Maravelias, 2013c).

Numerous effective commercial solvers are currently available to solve complex MILP models in a reasonable CPU time. Some of the most used ones are CPLEX, XPRESS and Gurobi. All of them include sophisticated search algorithms and contain wide range of pre-solving techniques, different types of cutting planes, heuristic techniques for finding feasible solutions, and algorithmic options to exploit parallel computing resources. The developments of efficient scheduling models together with the proper selection and configuration of these methods allow solving complex real-world scheduling problems in reasonable times.

MINLP Methods

The most basic form of an MINLP problem when represented in algebraic form is as follows (Grossmann, 2002):

$$\min Z = f(x, y)$$
$$s.t.\, g_j(x, y) \le 0\; j \in J \quad (MINLP)$$
$$x \in X,\, y \in Y$$

where $f\colon Rn \quad R^1$, $g\colon Rn \quad Rm$ are *differentiable* functions, J is the index set of constraints, and x and y are the continuous and discrete variables, respectively. In the general case the MINLP problem will also involve non-convex equations, which we omit here for convenience in the presentation. The set X commonly corresponds to a convex compact set, e.g. $X = \{x|x \in \mathbf{R}n,\, Dx \le d,\, xL \le x \le xU\}$ the discrete set Y corresponds to a polyhedral set of integer points, $Y = \{y|y \in \mathbf{Z}m,\, Ay \le a$, which in most applications is restricted to 0–1 values, $y \in \{0, 1\}m$. In most applications of interest the objective and constraint functions $f(\cdot)$, $g(\cdot)$ are linear in y (e.g. fixed-cost charges and mixed-logic constraints): $f(x, y) = cTy + r(x)$, $g(x,y) = By + h(x)$. The derivation of most methods for MINLP assumes that the functions f and g are convex.

In contrast to MILP, there is greater variety of solution methods and computer codes for solving MINLP problems. A review of methods that have addressed the solution of problem (MINLP) can be found in Grossmann (2002), Bonami et al. (2008), and in the website http://www.minlp.org. These methods include first branch and bound methods (BB), which are generally a direct extension of the LP-based branch and bound method for MILP, except that in this case an NLP solver (e.g. reduced gradient, successive quadratic programming, or interior point method) is used at each node, while the use of cuts at this point is still rather limited. A second class of methods includes decomposition algorithms such as Generalized Benders Decomposition (GBD) and Outer-Approximation (OA), in which there is an iterative sequence of NLP subproblems with fixed 0–1 variables that yield upper bounds, and MILP master problems that yield lower bounds. Convergence in this case is achieved when bounds lie within a specified tolerance. In order to avoid the repeated sequence of MILP master problems, branch and cut methods (or LP/NLP based branch and bound) have been proposed in which a single branch and bound tree is enumerated by updating the linear approximations with solution of NLP subproblems at selected nodes. Finally, a variant of the decomposition methods is the extended cutting plane method (ECP) in which the NLP subproblem is replaced by function evaluations, with which the algorithm effectively reduces to a successive MILP method.

The number of computer codes for solving MINLP problems has increased in the last decade. The program DICOPT (Viswanathan & Grossmann, 1990) is an MINLP solver that is based on the outer-approximation algorithm with some extensions, and is available in the modeling system GAMS (Brooke, Kendrick, & Meeraus, 1988). A similar code to DICOPT, AAOA, is available in AIMMS. Codes that implement the branch-and-bound method solving NLP subproblems at each node include the code MINLP_BB that is based on an SQP algorithm (Leyffer, 2001) and is available in AMPL, and the code SBB which is available in GAMS (Brooke et al., 1988). The code -ECP that is available in GAMS implements the extended cutting plane method by Westerlund and Pettersson (1995), including the extension by Westerlund and Pörn (2002). The code MINOPT (Schweiger and Floudas, 1998) also implements the OA and GBD methods, and applies them to mixed-integer dynamic optimization problems. The open source code Bonmin (Bonami et al., 2008) implements the branch and bound method, the outer-approximation and an extension of the LP/NLP based branch and bound method in one single framework. FilMINT (Abhishek, Linderoth, & Leyffer, 2010) also implements a variant of the LP/NLP based branch and bound method. Codes for the global optimization that implement the spatial branch and bound method include BARON (Sahinidis, 1996), LINDOGlobal (Lindo Systems, 2010), Couenne (Belotti, Lee, Liberti, Margot, & Wächter, 2009), and GloMIQO (Misener & Floudas, 2013).

It should be noted that solving MINLP problems is still often a non-trivial task, especially when compared to solving MILP problems. A major reason for this is the need to solve NLP subproblems, which often require good initial points to ensure convergence. Therefore, often an approach that is used in scheduling is to reformulate the problem as an MILP by for instance using exact linearizations of products of binary and continuous variables, or using piecewise linear approximations. This, however, can only be done in special cases.

Constraint Programming

Constraint programming (CP) is a relatively young field that grew out of the logic programming and computer science communities. It is not yet as broadly applied as mathematical programming, but it has achieved some of its most notable successes in scheduling and related areas. Industrial applications of CP include supply chain management, staff scheduling, airline crew rostering, container port scheduling, assembly line sequencing, electric grid management, water system control, computer processor scheduling, gate allocation in airports, and machine control in manufacturing. Applications specifically in process industries include production sequencing, lot sizing, processing network design, coordination of production and packaging with buffer tanks, and oil tanker berth scheduling (Malapert et al., 2012, Novas and Henning, 2010, Novas and Henning, 2012 and Zeballos et al., 2011). A wide variety of company-specific applications are detailed in Simonis (2001). Much of CP solution technology is summarized in a handbook (Rossi, van Beek, & Walsh, 2006), as are applications in scheduling, planning, vehicle routing, configuration, and networks.

As the name suggests, CP formulates a problem by writing constraints, but unlike MILP solvers, the CP solver processes the constraints sequentially to reduce the space of possible solutions. The reduced space is passed from one constraint to the next, which reduces it further, in a process known as *constraint propagation*. This entire process is carried out at each node of a search tree. Another major difference from MILP is that problems are typically written with a few highly structured *global constraints* that capture common modeling situations, rather than with many elementary constraints, such as inequalities or equations (Régin, 2011). More elementary side constraints can be added if necessary to fit a specific modeling situation. Global constraints are processed with *filtering* algorithms that are specifically designed for each type of constraint. These algorithms operate on each variable domain (the set of values that the variable can take) by removing some or all values that cannot occur in any feasible solution. This reduces the number of alternatives that must be examined in a tree search. An extensive catalog of global constraints can be found in Beldiceanu, Carlsson, and Rampon (2010) and a more compact catalog inHooker (2012).

One of the advantages of CP is that global constraints reveal problem substructure to the solver that might be missed by other types of solvers. Complicating side constraints tend to make a problem easier rather than harder, because the additional constraints result in more filtering and propagation. CP methods were originally conceived as a tool for finding feasible solutions, and they are sometimes less efficient for cost minimization. However, all state-of-the-art solvers can find optimal solutions.

State-of-the-art CP solvers include the open-source solver Gecode (Schulte & Tack, 2010); the IBM/ILOG CP Solver and Scheduler, which is invoked by the OPL Studio modeling system (van Hentenryck, Lustig, Michel, & Puget, 1999); Cosytec's CHIP solver (Simonis, 1996); NICTA's G12 solver (Stuckey et al., 2005) and OR-tools by Google. CP solvers tend to be very effective up to a certain problem size, but performance may degrade substantially for larger instances, or when the problem is perturbed slightly. Methods that combine CP with other optimization techniques tend to be more robust (Section 5.5). Constraint programming is further explained in the Supporting Content.

Heuristic and Metaheuristic Methods

Rule-Based Scheduling

The simplest realizations of scheduling algorithms are dispatching rules. Dispatching rules are constructive heuristics by which a schedule is built. In their simplest form, they order the queue of jobs that are waiting in front of a resource according to some, usually local, criterion, and assign the highest priority job to the resource when it becomes available. A huge number of dispatching rules have been proposed and compared in the literature on job shop scheduling. Panwalkar and Iskander (1977) already listed 113 dispatching rules and 36 papers in this domain. Following Haupt (1989), the dispatching rules use the following types of attributes to compute the priority of a job: arrival times (as FCFS, first come first serve), processing time parameters (as SPT, schedule the job with the shortest processing time on this station among those that wait in front of a station), static due date related criteria (e.g. EDD, earliest due date first, or ODD, earliest operation due date first), dynamic due-date related criteria (variants of measures of the slack, i.e. the time buffer remaining to process the job) and information about the queue in front of the station that the job has to visit next (e.g. the total processing time of this queue). A large number of simulation studies have confirmed that the shortest processing time first (SPT) rule provides the least mean flow time under almost all conditions among all simple dispatching rules. Somewhat surprisingly, it also gives very good results for the mean tardiness especially in situations where the load is high so that many jobs are finished late. This generalizes the rigorous result (Baker, 1974) that the SPT rule minimizes the mean flow time as well as the mean lateness in a single machine problem when all starting times are equal. The disadvantage of the SPT rule is that, obviously, under this rule some long jobs incur large delays, consequently the maximum lateness and the variance of the lateness become large. With respect to the maximum lateness, due date related criteria are preferable. Among the simple priority rules, probably the best compromise is the CR + SPT rule (Anderson & Nyirendra, 1990) that combines a slack-based criterion for jobs that are not beyond the due dates with the SPT rule for the inevitably late jobs (Moser and Engell, 1992 and Pierreval and Mebarki, 1997). Several attempts have been made to introduce a dynamic selection of dispatching rules (e.g. Pierreval & Mebarki, 1997) but the improvements reported are moderate. An obvious drawback of scheduling rules is their myopic nature: only the local situation in the queue considered is taken into account but not the situation that the job will encounter during the future processing steps. Possible extensions are to include the situation in the next queue (see e.g. Holthaus & Rajendran, 2000), to include a limited look-ahead (Moser, Herrmann, & Engell, 1992) or to perform a limited search (beam search, Sabuncuoglu & Bayiz, 1999, arc greedy heuristics, Pranzo, Meloni, & Pacciarelli, 2003).

Decomposition-Based Approaches

As observed by many authors (e.g. Kelly & Zyngier, 2008), solution performance is crucial for real-world applications of advanced scheduling methods and optimal solutions often cannot be computed for large problems within reasonable timeframes. Simplification and decomposition are the obvious remedies to render real-world problems computationally tractable. Common approaches to problem decomposition include hierarchical, time-based, resource-based, and order-based approaches (Bassett et al., 1996,Dimitriadis et al., 1997 and Kelly and Zyngier, 2008).

In time-based decomposition, the problem at hand is solved for a subset of the tasks that can be started or are due to be finished within a certain horizon. Then the first scheduling decisions are implemented (or fixed in simulations) and after a certain period, the problem is solved again, taking into account the present state of the execution of the schedule, resource availability and orders. This is similar to the moving-window approach of model predictive control. Of course the limited look-ahead of this approach leads to a certain amount of sub-optimality, but on the other hand it is assured that the decisions that have to be taken within the immediate future are indeed computed in real-time.

Another obvious idea to reduce the combinatorial complexity of scheduling problems is to focus on the scheduling of those resources that are critical. The overall flow time of the jobs is hardly influenced by the processing on stations with a low utilization of their capacity, so it is not mandatory to include them into a rigorous optimization. On the other hand, in systems with a high utilization of the capacity, the position of the bottlenecks will vary dynamically and the problem cannot be reduced a priori to the scheduling of a few key resources. The advantage of focusing on the bottlenecks is that for these, theoretical results and efficient algorithms for one-machine problems can be utilized. An efficient algorithm that combines a "bottleneck first" heuristic with rigorous optimization was proposed by Balas, Lenstra, and Vazacopoulos (1995). The algorithm sequences the resources sequentially by solving single resource problems, keeping the previously determined sequences on the other stations fixed and ignoring the constraints on the not yet scheduled resources. The bottleneck among the remaining resources is determined by solving the single machine problems for all not yet scheduled resources. After the jobs have been scheduled on all resources, the stations are rescheduled one after the other by the same approach for a predefined number of rounds. The solution of the single resource problems is rigorous, making use of the results of Carlier (1982) and Potts (1980). It turned out that this is an approach that is hard to beat (Pranzo et al., 2003).

Thirdly, iterative constructive building and refining of schedules can be used to generate close-to-optimal schedules within short computation times (e.g. Kopanos et al., 2010). Here orders are inserted into a schedule one-by-one, fixing allocation and sequencing decisions for previously inserted orders. Once a first feasible schedule has been obtained, refinement techniques (local search methods or partial re-optimization) can be applied to improve the solution quality. Obviously, this approach fits well into a moving horizon scheme; the construction process considers the orders within the prediction window and the refinement can be executed until the next decision must be taken (e.g. a job frees a resource or the starting time of a new one has been reached).

Meta-Heuristics

The fact that many scheduling problems are hard to solve by exact algorithms, especially if very detailed models of the problem are required to check feasibility and optimality, has motivated the application of meta-heuristic techniques. Meta-heuristics are characterized by a guided stochastic search. Due to the stochastic nature of the search, meta-heuristics can escape from local minima in non-convex nonlinear problems or in problems with disconnected feasible regions. Most meta-heuristics mimic processes in the nature:

- Simulated annealing (SA) is derived from the formation of defect-free crystals in the cooling of melts
- Evolutionary algorithms are derived from the evolution of species in nature
- Ant or bee colony optimization is derived from the search for food by animals that live in large populations
- Particle swarm optimization is derived from the behavior of swarms of animals with communication between the members.

All meta-heuristics try to improve the current solution or the current best solution within a set of solutions by performing random steps in the space of the free parameters, so they are *improvement* heuristics. Depending on the size of the steps, the solution space is explored broadly (exploration) or locally (exploitation, refinement). In population-based algorithms (all of the above except SA), a set of possible solutions are modified in each iteration. The parameterization of the algorithms defines the transition from global to local search or the switching between the two mechanisms. The algorithms usually have to be tuned carefully for the problem at hand. Locally, the search usually is less efficient than gradient-based algorithms, and for discrete problems it becomes an exhaustive local search which is terminated if the progress becomes too slow or a predefined number of iterations is reached. Meta-heuristics usually do not provide the optimum for the problem at hand (and if they do so, there is no guarantee that the solution is indeed the global optimum) but often lead to good solutions that satisfy the constraints within a limited amount of time.

An attractive feature of meta-heuristics compared to mathematical programming is that they can be easily integrated with sophisticated simulation models that represent many details of the problem at hand, including features and constraints that cannot easily be modeled by equations or that would give rise to models with a large number of variables but relatively few free decision variables. Often such simulation models are built in industry with considerable efforts for parameter studies and for the analysis of the effect of modifications, and the effort to reformulate them as equation-based models would be uneconomic, if at all possible.

A difficult issue in the application of meta-heuristics in scheduling problems is the representation of constraints. Some constraints can be mapped into the problem representation and into the choice and parameterization of the stochastic operators that generate new solutions. For example, the fact that a valid schedule has to be a permutation of the sequence of jobs or operations under consideration can be represented by manipulating a string of integers of suitable length where each integer represents one operation, and where the changes are restricted to permutations on this string. Such an implicit representation is usually only possible for some basic constraints of the problem at hand. If additional constraints are present, usually many solutions are generated that violate some of them. The simplest approach is to discard all solutions that violate one of them. This strategy is efficient if the detection of the infeasibility is computationally much cheaper than the evaluation of a proposed solution with respect to the solution quality. In order to direct the search into promising regions of the search space, usually penalty functions are used (indirect representation of the constraints). If, for a minimization problem, the penalty of an infeasible solution is always larger than the worst value of the cost function of a feasible solution, feasible solutions are always preferred over infeasible ones, and a carefully chosen penalty function steers the solution process in the direction of the feasible subregion(s). On the other hand, a large penalty on infeasible solution may prevent the search process from traversing infeasible sub-regions in disjoint search spaces and therefore the search may get stuck into suboptimal sub-regions (Sand et al., 2008).

Other frequently used constraint handling strategies include the use of repair mechanisms by which infeasible solutions are turned into feasible ones during the search process (i.e. a candidate solution is modified by local improvement algorithms such that a feasible schedule results); or the handling of only some of the degrees of freedom by the meta-heuristic search strategy and fixing the remaining ones during the evaluation of the solution, e.g. by using local priority rules (cf. Piana & Engell, 2011).

Heuristics-Guided Exhaustive Search by Reachability Analysis of Timed Automata

While in MILP formulations, the mathematical formulation of the optimization problem is embedded in a search space with a dimension that is higher than the number of degrees of freedom (decisions) of the original problem in order to transform it to a set of linear constraints, in the approach based on *timed automata* (Alur & Dill, 1992), the set of all feasible schedules is described implicitly by expressing the constraints on the schedules (precedence constraints, exclusive use of resources, timing constraints, for instance, zero delay or ripening times, and material balances) by means of timed automata and shared variables (Fehnker, 1999 and Abdeddaim and Maler, 2001). Using timed automata, the sequence of operations that have to be performed for each job or batch and the conditions for the use of resources can be modeled in a modular fashion. The individual automata are composed to a global automaton. Every run through the locations of the global automaton from the initial location where all operations are waiting to start their execution to the final target location where all recipe operations have finished their execution represents a feasible production schedule. TA models can be generated automatically from RTN descriptions of the scheduling problem. The global timed-automaton model is searched for the best schedule using (best-cost) reachability analysis (Larsen et al., 2001, Abdeddaim and Maler, 2001 and Panek et al., 2006). If the search terminates within the available computational resources (computing time and memory), the optimum is found, otherwise the best schedule found within these constraints is returned.

The TA-approach has been applied to benchmark job shop problems and the solution efficiency compared favorably to other techniques (Abdeddaim and Maler, 2001 and Panek et al., 2006). In (Panek, Engell, Subbiah, & Stursberg, 2008), TA-based scheduling was compared to state-of-the-art MILP techniques for a variant of the Westenberger and Kallrath benchmark problem (Kallrath, 2002) and the performance was comparable or even superior. Modular modeling by TA and computation of schedules by reachability analysis can be extended to problems with intermediate due dates and minimization of tardiness (Subbiah, Tometzki, Engell, & Panek, 2009) and to problems with sequence-dependent changeover times (Subbiah, Schoppmeyer, & Engell, 2011). Problems with significant continuous degrees of freedom, as e.g. the choice of batch sizes, can only approximately be modeled by TA. The strength of the TA-based approach is that it provides very good schedules early in the search process, hence within short computation times. Therefore, it is most suitable for problems which cannot be solved by exact methods within the time frames demanded by the application. Another advantage is that the modeling is modular and graphical and hence intuitive. For details see the supplementary material.

Hybrid Methods

Optimization methods frequently have complementary strengths that can be combined in *hybrid methods*. These include combinations of two or more exact methods, as well as combinations of exact and heuristic methods.

The best developed combinations of exact methods integrate constraint programming (CP) and mathematical programming (MP). One scheme combines propagation from CP and relaxation from MP in a branching search (Bockmayr and Kasper, 1998, Bockmayr and Kasper, 2004 and Hooker and Osorio, 1999). This can be particularly effective for scheduling problems, in which constraints frequently have weak relaxations but propagate well, or vice versa. A second scheme is a *branch-and-price* algorithm in which CP methods solve the pricing subproblem (Junker et al., 1999 and Easton et al., 2004). An advantage of this approach is CP's ability to deal with the many complicating, combinatorial constraints that tend to appear in the subproblem. It can be useful for process scheduling when there are a huge number of ways to route and schedule batches along paths through a processing network. A number of CP-based constraints are useful for describing feasible paths and schedules.

A third scheme for uniting CP and MP is through *logic-based Benders decomposition* (Hooker and Ottosson, 2003 and Hooker and Yan, 1995). In the scheduling area, the idea is most naturally applied to problems with both an assignment and a scheduling component. For example, one may wish to assign batches to facilities for processing, and then schedule the batches assigned to each facility, perhaps with the objective of minimizing makespan (Jain & Grossmann, 2001). If the cuts are properly designed, an approach of this kind can reduce solution time by several orders of magnitude (Hooker, 2006 and Hooker, 2007a). CP/MP hybrids are further discussed in the Supporting Content and surveyed in Hooker (2012). Software packages such as ECLiPSe (Wallace et al., 1997 and Rodosek et al., 1999), OPL Studio (van Hentenryck et al., 1999), Mosel (Colombani & Heipcke, 2002), G12 (Stuckey et al., 2005), SIMPL (Aron, Hooker, & Yunes, 2004, Yunes, Aron, & Hooker, 2010), SCIP (Achterberg, 2008), and Comet (Michel, See, & Van Hentenryck, 2009) implement CP/MP hybrid methods to some extent. MIP/CP hybrid methods have been proposed to address scheduling problems both in sequential (Harjunkoski & Grossmann, 2002) and network (Maravelias and Grossmann, 2004 and Roe et al., 2005) chemical production environments.

Exact solution methods can also be advantageously combined with heuristic methods. One approach is to apply exact methods to subproblems with a limited number of degrees of freedom and a mathematical structure that is amenable to efficient rigorous algorithms (in particular MILP solvers), while using heuristic methods to deal with non-convex constraints or to couple the subproblems. A hybrid method combining a MIP model with a heuristic scheduling algorithm was propped for the scheduling of multi-stage processes (Maravelias, 2006).

An example of subproblem coupling occurs in the solution of *two-stage stochastic programming* problems in which uncertainties are represented by a set of scenarios. The first-stage decisions are identical for all scenarios, whereas the remaining second-stage or recourse decisions are adapted to the parameters of the individual scenarios. The goal is to optimize the expected value of the cost function over all scenarios for optimal choices of the scenario-specific second-stage decisions. When the first-stage variables are fixed, the scenario sub-problems decompose and can be solved by exact methods. Till, Sand, Urselmann, and Engell (2007) proposed a hybrid algorithm in which the first stage variables are handled by an evolutionary algorithm and the subproblems are solved by MILP techniques. The algorithm scales almost linearly in the number of subproblems and generates solutions very close to the solutions provided by using an exact method for the full problem (Tometzki & Engell, 2009). A significant advantage of this approach is that by using population-based algorithms, Pareto fronts can be computed for multiple targets (in this case the expected profit and a risk function) at almost no additional computational cost (Tometzki & Engell, 2011).

Optimization can also be combined with local search, as for example in *large neighborhood search* (Pisinger and Ropke, 2010 and Shaw, 1998), sometimes called *very large neighborhood search* (Ahuja, Ergun, Orlin, & Punnen, 2002). The simplest approach is to unfix some of the variables in the current solution of a local search algorithm and use optimization methods to find the best solution in the resulting large neighborhood of solutions. Branch and bound, CP methods, and dynamic programming have been used for his purpose. Another possibility is to define the neighborhood in such a way that it can be searched by a polynomial-time algorithm, such as a matching or shortest-path algorithm. Sometimes the neighborhood size is increased dynamically when its best solution is no better than the current one. Hybrids of local search and CP are surveyed by Shaw (2011), and hybrid metaheuristics by Blum, Puchinger, Raidl, and Roli (2011). A number of techniques introduce local search into branch-and-bound methods, such as local branching (Fischetti & Lodi, 2008) and feasibility pumps (Fischetti, Glover, & Lodi, 2005).

Rescheduling

Scheduling is an inherently dynamic process with new information becoming available continuously and, in addition, multiple sources of uncertainty. To address these two issues, *rescheduling* should be performed iteratively, at latest before the end of the current scheduling horizon, which means that, in a dynamic environment, any scheduling activity is in essence rescheduling. While rescheduling has been viewed primarily as a method to address uncertainty (e.g. demand uncertainty), it is important to note that rescheduling is necessary also to account for new information becoming available (e.g. arrival of new orders) as the planning horizon moves forward. There are two major approaches to rescheduling: deterministic and stochastic. In the former, a deterministic model is solved at each iteration to *react* to the new information and the realization of uncertainty, while in the latter a stochastic problem is formulated to account for (at least some) uncertain parameters.

Deterministic reactive rescheduling is the method which is most similar to what human schedulers naturally do. Once an event that triggers rescheduling occurs the existing schedule is modified so that feasibility is restored and quality degradation is reduced. Deterministic approaches can be further classified based on how the model is updated to reflect the current state of the system and how the current schedule is modified. In general, upon observing a disturbance, part of the current schedule is fixed and the remainder of the scheduling horizon is re-optimized. Such approaches were proposed by Méndez and Cerda (2003b),Relvas, Matos, Barbosa-Povoa, and Fialho (2007), and Novas and Henning (2010). Also, Rodrigues, Gimeno, Passos, and Campos (1996) proposed a rolling horizon reactive scheduling method in which they provide a predictive framework to determine future infeasibilities, while van den Heever and Grossmann (2003) proposed a receding horizon approach for rescheduling a hydrogen supply chain. Kopanos, Capon-Garcia, Espuna, & Puigjaner (2008) addressed the inclusion of costs for modifications of the initial schedule defined by the rescheduling in order to preserve the production plant stability and avoid the generation of pseudo optimal schedules. Finally, Subramanian, Maravelias, and Rawlings (2012) recently proposed a state-space scheduling formulation, which allows explicit modeling of disturbances and thereby an automatic formulation of the rescheduling model.

Stochastic optimization approaches attempt to explicitly model the uncertainty in the system and generate schedules that account for this uncertainty. Typical sources of uncertainty include customer demands, quality of products and intermediates, supply of raw materials, and processing times, all of which can be modeled as stochastic parameters. However, it is important to stress that there are stochastic events, such as unit breakdowns, which can lead to rescheduling but cannot be readily modeled as stochastic parameters. Given the wide range of approaches to general optimization problems under uncertainty (Birge & Louveaux, 1997; Kall and Wallace, Sahinidis, 2004), there is a variety of stochastic approaches to rescheduling, including specialized search algorithms (Balasubramanian and Grossmann (2002), fuzzy programming (Balasubramanian & Grossmann, 2003), simulation-based optimization (Honkomp, Mockus, & Reklaitis, 1999), multi-parametric methods (Li and Ierapetritou, 2008a, Li and Ierapetritou, 2008b and Ryu et al., 2007), robust optimization (Janak et al., 2007 and

Lin et al., 2004), and two-stage stochastic programming approaches (Cui and Engell, 2010 and Sand and Engell, 2004).

Clearly, the advantage of approaches that account for uncertainty is that they can lead to better solutions if solved effectively. However, the modeling and evaluation of the impact of uncertainty typically leads to computationally expensive approaches which, at the present stage, cannot be used to address industrially relevant short-term planning problems (there are no industrial applications known to the authors). For example, the solution of multi-stage stochastic programming models with discrete recourse remains a challenging problem despite the development of specialized methods (see e.g. Sand et al., 2008 and Till et al., 2007). Multi-parametric programming approaches attempt to overcome this limitation by off-line computations, which nevertheless can also become prohibitively expensive. Robust optimization (Ben-Tal, El Ghaoui, & Nemirovski, 2009) is the most attractive in terms of computational effort, but it leads to conservative schedules.

Deterministic approaches, on the other hand, can lead to solutions that are suboptimal or even infeasible when evaluated under uncertainty (this may also be true for optimization methods under uncertainty), but are computationally more effective, which has three important implications, as discussed in Subramanian et al. (2012):

- They can be solved to (near) optimality in reasonable time, providing solutions that are better, even when evaluated under uncertainty, than solutions obtained within the same time frame using approaches that account for uncertainty.
- They allow us to reschedule more frequently, thus reacting faster to changes and new information, thereby resulting in better implemented (closed-loop) schedules.
- They enable us to consider longer horizons which often is as important as accounting for uncertainty; e.g. considering future demand spikes can result in increased production in the near term.

Furthermore, deterministic methods are more natural for short-term scheduling problems where (i) solutions have to be found quickly (short response time), (ii) the planning horizon is short and thus the uncertainty limited, and (iii) feasibility is often more important than optimality because problems are highly constrained. Methods that account for uncertainty are more reasonable in problems where (i) decisions are made infrequently, so stochastic approaches can be solved to (near) optimality, (ii) the planning horizon is long and thus uncertainty is higher, and (iii) there are many optimization degrees of freedom.

Finally, it is important to note that regardless of the method, a number of challenges have to be overcome in order to implement effective rescheduling in practice. For example, it is necessary to know the current plant state and the already committed resources (see Méndez, Cerdá, et al., 2006). Also, a major hurdle is that no general rescheduling framework for a large class of industrial scheduling problems is available. The state-space framework of Subramanian et al. (2012) addresses some of these challenges, including the modeling of the state of the system, the explicit modeling of events that lead to rescheduling, and the automatic formulation of the re-optimization problem solved at each iteration.

In general, deterministic rescheduling approaches are preferred in industrial practice due to their intuitive realization. They include simple heuristic algorithms which typically rely on different kinds of "buffers". The available production capacity is not fully used in the nominal case, and intermediates and final products are stored to compensate for production unit breakdowns. In the case of uncertain events such as bad product quality, rush orders or equipment breakdowns, these buffers can be used to react without violating constraints. However, buffers lead to higher costs, such that the operational strategy is a compromise between profit maximization and risk minimization. In abstract terms, the plant is operated away from its constraints at the price of reduced benefit.

PRODUCTION SCHEDULING SOLUTIONS

Having explored various aspects and approaches for production scheduling it is of great interest and also one focus of this paper to discuss the application of those in an industrial environment. In this section existing software is shortly discussed together with some selection criteria on both models and methods.

General vs. Specific approaches

A major issue that arises in the development of models for process scheduling is the relative merit of general vs. specific models to production scheduling (e.g. see Méndez, Cerdá, et al., 2006). Ideally, one would of course like to develop models that are as general as possible, and implement them in general purpose software packages, similarly as is done in process simulators (e.g. Aspen Plus, Hysys), as this would presumably provide powerful tools for addressing a greater variety of scheduling problems. For instance, being able to address plants of arbitrary configuration, with and without intermediate storage, with possibility of splitting and mixing batches, accounting for fixed and variable processing times, sequence dependent changeovers, and resource constraints for utilities and manpower, demands specified at arbitrary time points or as constant demand rates, would be highly desirable. In fact, MILP models accounting for material balances based on a common time grid that have been proposed for short term batch scheduling come quite close to satisfying these requirements.

Unfortunately, since scheduling problems are NP hard, the efficient solution of their corresponding MILP models can become computationally very expensive for large-scale problems. Furthermore, it is often the case that scheduling problems exhibit special features such as multistage plant structure, fixed batch sizes and fixed processing times which can greatly simplify the mathematical formulation, which in turn increases the likelihood for the solution of large-scale industrial problems. A specific example is the scheduling of multistage batch plants with single unit per stage, zero-wait policy and sequence-dependent changeovers, which can be formulated as an asymmetric traveling salesman problem (Pekny, Miller, & McRaec, 1990), which can be solved much more effectively than for instance through an STN-MILP model.

In summary, the major dilemma one often faces in production scheduling is on the one hand developing general purpose models that can cover more problem classes but at a computational cost that is likely to be high, which can render its application to industrial problems to be rather limited. On the other hand, developing special purpose models that exploit the special structure of the scheduling problem can lead to more efficient computational times, but at the expense of having to develop tailored models. Presumably this expense can become larger than developing the one single generic model, much in the same spirit as is done in process simulation where generic packages such as ASPEN Plus and Hysys are used in place of specific tailored models.

It is our belief that while it is always worth to aim for generic models, in production scheduling it pays on balance to develop specific models that can exploit the particular structure of the problem. Major reasons for this statement are that unlike process simulation where a major component is thermodynamic data that can be shared across many applications, the data for scheduling are specific to each application. Furthermore, as opposed to process simulation where models tend to be procedural, process scheduling models are equation-based and can be implemented without great expense and time investment in modeling systems described in the following Section 6.2. These packages have greatly facilitated the development of mostly MILP and less frequent MINLP and CP models for process scheduling.

In conclusion, we believe that the proper role of generic models is to treat them as a generalized case of specific models. Generic models should be used when many features must be simultaneously accounted for in a given problem. Otherwise, when not all the features are present, it is often more effective to exploit this fact with specific models to produce much more efficient models and solution methods. Consequently, the idea of developing generic monolithic software for scheduling similarly as in process simulation is most likely to fail. A much more fruitful approach is to take advantage of modern modeling software systems so as to develop tailored scheduling models that exploit the structure of the corresponding problem and in that way effectively address industrial sized problems.

Commercial Modeling Systems

As described in the previous sections, the most common form of scheduling models are MILP models, or MINLP models when it is necessary to explicitly model nonlinearities related to cyclic schedules, inventories or process models. In either case, it is possible to implement these models in commercial modeling systems since the models are expressed in equation form. This feature represents a significant advantage when compared for instance to process simulation, in which procedural approaches are common in large part due to the need of handling thermodynamic properties. Therefore, it is fair to say that most scheduling models have been implemented in commercial modeling systems like AIMMS, AMPL, GAMS and OPL. Furthermore, the effort involved is relatively modest.

All modeling systems these days allow the user to implement models in the form of algebraic models involving discrete and continuous variables and constraints and an objective function. Since it is possible to use indexed variables and constraints it is easy to implement models in compact form, similar to those expressed analytically, with which large-scale problems are generated automatically by appropriate definition of sets for the indices (e.g. products, lines, time intervals) and corresponding data for the indexed parameters. Furthermore, a major advantage of these modeling systems is that they automatically interface with optimization solvers and the users need not be concerned with low level programming details (e.g. MPS format, or supplying matrices and calling subroutines) to activate the optimization solvers. Moreover, for nonlinear models, these modeling systems perform automatic differentiation, the user need not be concerned with supplying derivative information for the nonlinear solvers.

Finally, the modeling systems have also to a varying degree capability of interfacing with spreadsheets, databases or graphics packages, thereby facilitating the deployment of the model as a tool with graphical user interface that can be used by non-specialist users. Using these features, it is very easy to create new test instances and demonstrate the potential of optimization, as well as embed the solution into e.g. MS-Office environment as the mathematical models are automatically propagated based on the given problem data. Even if the building of such a system might require deeper technical skills, by utilizing available interfacing tools, the optimization solution can be easily managed and even slightly configured by the non-expert. The only drawbacks are the Software (SW) dependencies and consequently risk of multiple upgrade needs, as well as potentially high licensing costs of several components. AIMMS has additionally graphical capabilities that allow easy implementation of prototype applications including Gantt chart displays, which in some cases may be sufficient for a final optimization solution. Other modeling environments can be connected to various visualization components, which often involve separate licensing costs.

Lastly, one important aspect is that models in these platforms have upward compatibility and are often platform independent (Windows, Linux, Unix, ...) which ensures their long term use, especially if they are properly documented. However, the compatibility across the modeling systems is restricted, as no detailed common modeling language standards exist. Many of the modeling systems are text and file based, which may make them vulnerable to intellectual property leak. Nevertheless, the main benefit of using a professional modeling system is the possibility to efficiently develop, refine and evaluate mathematical modeling alternatives.

Commercial Operational Scheduling Software

The usual options are to either develop scheduling software in house or license a commercial scheduling software package. Nevertheless, the first question that needs to be answered is, if there is a true need to apply operational scheduling software or not. Fig. 6.1 shows one example of evaluating the necessary level of scheduling. If the statements in the lower rectangle are true, this indicates the need of a more advanced scheduling application. For example, if the manufacturing has a high utilization AND has shared resources AND has a complex product mix, then application of operational scheduling is necessary to reduce operating and capital expenditure.

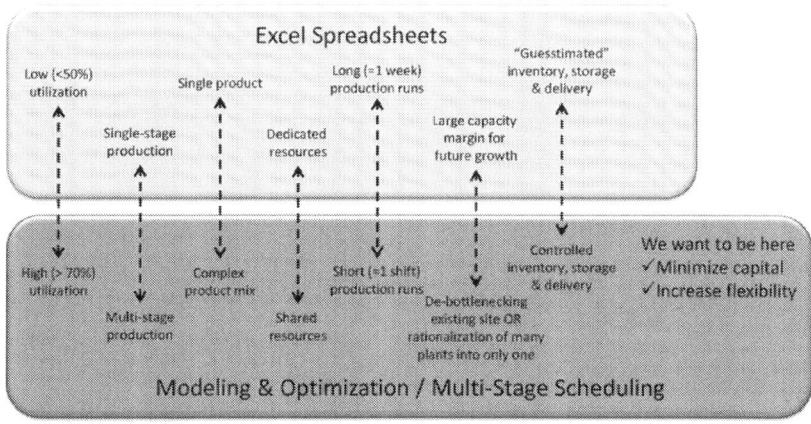

Figure 6.1: Is there a need for operational scheduling or not?

The main characteristics of the operations in Fig. 6.1 are:

- *Utilization*. If the manufacturing asset utilization is low, there are many possibilities to manufacture the products, which will hardly affect the efficiency.
- *Resources*. As shown in Fig. 2.3, there are various configurations of the key unit operations that may vary between products. A higher number of shared resources and stages needed make the scheduling problem more difficult.
- *Product mix*. Multiple products may give rise to change-over issues, which make the scheduling problem more complex.
- *Run length*. When the run length of the products is shortened, more products will be produced per time unit, leading to more batch operations to be optimized.
- *Control of the operation*. Knowing more exactly what is going on in the operations on a "real-time" basis, provides a solid base for scheduling.

When operational scheduling is necessary, there is abundance of commercial software available, among others from Advanced Process Combinatorics, Aspentech, Asprova APS, Infor, Intelligen, Manugistics, OMPartners, Optisol, ORSoft, Ortems, Preactor, Quintiq and Taylor. Most of these apply some heuristics, possibly mixed with mathematical optimization. The necessary elements of a solution discussed in the previous sections certainly apply to commercial tools as well.

Given the abundance of commercial scheduling software, which one to choose is a challenge. If the plant is using a business planning software (SAP, JDA, Oracle or others), the first choice is either to use the scheduling solution provided by the package, or to use the 'best in class' software. An additional challenge is that almost all companies who sell operational scheduling software will claim that their package can be used to schedule the plant under consideration. If it cannot be done directly, most companies claim that they can adjust some functionalities or adapt the tool such that their software can be used.

Selection of Methods and Models

As discussed in Section 6.3, the first question that needs to be answered is whether there is need for advanced production scheduling solutions, and if yes, whether existing software is sufficient. If an in-house solution is necessary, then a number of decisions have to be made, from the selection of processes and features to be modeled, to the selection of the modeling and solution approach. Next, we discuss these aspects and offer some general guidelines.

Problem Definition

The first decision concerns the *interaction* of the scheduling problem with the other planning functions and the types of decisions. For example, batching, assignment and sequencing decisions can be considered simultaneously; or batching decisions can be made prior to scheduling. Production in the process industries almost always involves resources other than processing units (e.g. cleaning-in-place, electricity), multiple processing stages, and tasks with multiple inputs and/or outputs. A scheduling model, however, has to account only for those constraints and characteristics that affect the feasibility or implementability of the solution. For example, if an additive is always available, then its addition does not have to be modeled. On the other hand, if production is regulated (e.g. through FDA), then the model should yield solutions that meet all regulatory requirements. In general, it is important to carefully select the boundary of the scheduling problem, and the constraints and characteristics to be modeled. The goal is to generate a model that is as simple as possible, so it can be solved effectively, ensuring that the obtained solutions can be transformed into implementable schedules.

Modeling and Solution Paradigm

Given a problem, one can develop scheduling approaches using different modeling paradigms and different solution methods for the resulting models. In general, there are six major types of modeling/solution options (see Table 6.1 for a summary): (a) scheduling algorithms (exact and heuristic), (b) general meta-heuristics (e.g. evolutionary algorithms), (c) timed automata, (d) integrated modeling and solution methods (e.g. CP), (e) mathematical programming, and (f) hybrid methods.

Table 6.1: Methods for problems in different production environments considering different sets of decisions. k = number of stages; $|Jk|$ = number of units in stage k; A: task-unit assignment; B: batching; S/T: sequencing/timing

Environment	Sequential			Network
	$k = 1, |Jk| = 1$	$k = 1, |Jk| > 1$		
	$k > 1, |Jk| = 1$	$k > 1, |Jk| > 1$		
Decisions	S/T	A, S/T	B, A, S/T	B, A, T
Methods				
Heuristics	√			
Meta-heuristics	√	√		
TA	√	√		
CP	√	√		
Math programming	√	√	√	√
Hybrid		√		√

In general, model-based methods become comparatively more effective as the complexity, but not necessarily the size, of the problem increases. If the selected method is model-based, then the user has to select a specific model. We present a critical review of the main types of models, with an emphasis on mathematical programming, and offer some guidelines for their selection. A summary of models that are available for different classes of problems is given in Table 6.2.

Table 6.2: Types of mathematical programming models for different classes of problems (based on processing characteristics and constraints)

Processing characteristics and constraints	Changeover times	Changeover costs	Storage restrictions	Shared utilities	Intermediate due dates	Variable batch sizes	Variable processing times	Holding and utility costs	Time varying utility costs
Sequential									
Precedence-based									
Immediate	✓	✓	✓		✓	✓	✓		
Global	✓		✓		✓	✓	✓		
Time-grid-based									
Discrete (common)			✓	✓	✓	✓		✓	✓
Continuous (unit-specific)	✓	✓			✓	✓	✓		
Network									
Time-grid-based									
Discrete									
Common	✓	✓	✓	✓	✓	✓		✓	✓
Specific	✓		✓		✓	✓		✓	✓
Continuous									
Common	✓	✓	✓	✓	✓	✓	✓		
Specific	✓	✓	✓		✓	✓	✓		

Sequential Production Environments

When the batching problem is solved independently, problems in sequential environments become similar to discrete manufacturing scheduling problems. Thus, a wide range of modeling options is available. Constraint programming is one choice, especially when the problem is tightly constrained or the goal is to find a feasible solution fast. TA-based approaches can also be used to address these problems effectively. If math programming models are employed, then the two main options are continuous-time precedence-based and time-grid-based models. Time-grid-based models tend to be tighter and computationally superior to precedence-based models, although small changes in the process configuration and objective function can affect the model performance drastically (Castro et al., 2006 and Mouret et al., 2011). However, precedence-based models are exact and easier to understand and no a priori decisions that affect solution quality, such as the number of time points, are needed. If batching decisions are included, then no scheduling algorithms are available and CP methods tend to be less effective, so the best choice is to use a MIP model. In general, time-grid-based models can represent more process characteristics. A common time grid is the only option when utility constraints have to be modeled.

Network Production Environments

The overwhelming majority of approaches for network environments are based on mathematical programming models. Since materials may be consumed and produced by tasks executed in multiple units and resources other than processing units should be taken into account, models that employ a common grid are the method of choice. If one or more of these characteristics are not present, then models featuring unit-specific grids, which are less general but computationally more effective, can be selectively employed. Discrete-time models can be easily extended, without increasing the computational effort, to linearly account for: inventory, holding, and utility costs; model intermediate due dates; and represent tasks with variable resource consumption during their execution. Recent studies on network production environments suggest that they are at least as efficient as continuous-time models (Sundaramoorthy & Maravelias, 2011a). However, in addition to the more accurate representation of events, continuous-time models can be easily customized to exploit problem features and are more effective in dealing with sequence-dependent changeovers. This means that if the structure of the problem is simple and other complicating constraints are not present (e.g. utility constraints), then a continuous-time model is likely to be more effective than a discrete-time model. Finally, time-grid-based models have been proposed to address problems in facilities that consist of sequential and network subsystems (Sundaramoorthy & Maravelias, 2011b) as well as facilities that involve blending operations (Kelly, 2006).

SUCCESSFUL APPLICATIONS FROM THE PROCESS INDUSTRIES

Aside from obvious confidentiality issues, finding trustworthy evidence of the exact impact of scheduling optimization is not as trivial as one could imagine due to the following reasons: Few companies have actually implemented fully automated systems without any manual steps; as the data is normally stored in various places, backtracking to see the real benefits is difficult; proper analysis tools are missing and often there are independent parallel improvement projects that also contribute to an increased productivity. Moreover, the key performance indicators (KPIs) needed for estimating benefits are not clearly defined and it is often also within the interest of companies not to reveal figures on real improvements, as it may have a negative impact on their businesses. In the following, few examples from the process industries are discussed shortly.

Example from the Dairy Industry

A benchmarking study from the dairy industry was presented in Bongers and Bakker (2006), where a 'simple' example system was described and is depicted in Fig. 7.1.

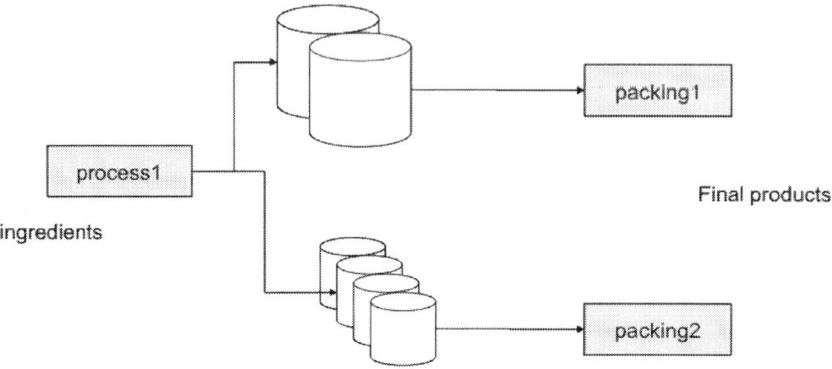

Figure 7.1: Example configuration to test commercial software packages.

The key characteristics of this example system are that it has close to 100% utilization and the optimal sequence for the packaging lines (when neglecting the process) is the worst sequence for the process (when neglecting the packaging). In the study, a number of software packages were tested, many of which were not able to deal with the full complexity of this 'simple' system. The most successful software package (INFOR) was tested on a more complex problem from ice cream manufacturing described in Bongers and Bakker (2008) with the following main properties:

- 8 packaging lines with different size and amount of associated buffer tanks.
- One processing line feeds all packaging lines.
- In total 130 stock keeping units (SKUs) and 160 recipes.
- Fresh dairy ingredients (limited shelf life) also restricting the buffering times.
- Stringent cleaning regime on process (Allergens & Kosher).
- Mandatory cleaning policy (24 h cycle on process, 72 h on other equipment).

The application of a scheduler resulted in an overall increase of approximately 30% of the factory output. The report in this industrial case study has created a rich set of literature and some of the successful approaches can be found in e.g. Kopanos, Puigjaner, and Georgiadis (2012), van Elzakker, Zondervan, Raikar, Grossmann, and Bongers (2012), and Subbiah et al. (2011).

Scheduling Optimization in the Petrochemical Industry

Optience Corporation has successfully implemented planning and scheduling applications based on NLP and MINLP models for Olefins and Aromatics businesses using their SCMart™ software since 2004 (seeFig. 7.2). The optimization problems possess a challenge due to the nonlinearity and discrete nature that arise from rigorous modeling approach to represent blending of raw materials, reaction in cracker and reformer as well as separation constraints. The complex optimization trade-off is much more pronounced when the process utilizes naphtha as the main raw materials. An implementation at Mitsubishi Chemical Corporation (MCC) resulted in optimal naphtha feedstock allocation and blending and a 2% improvement in their operation.

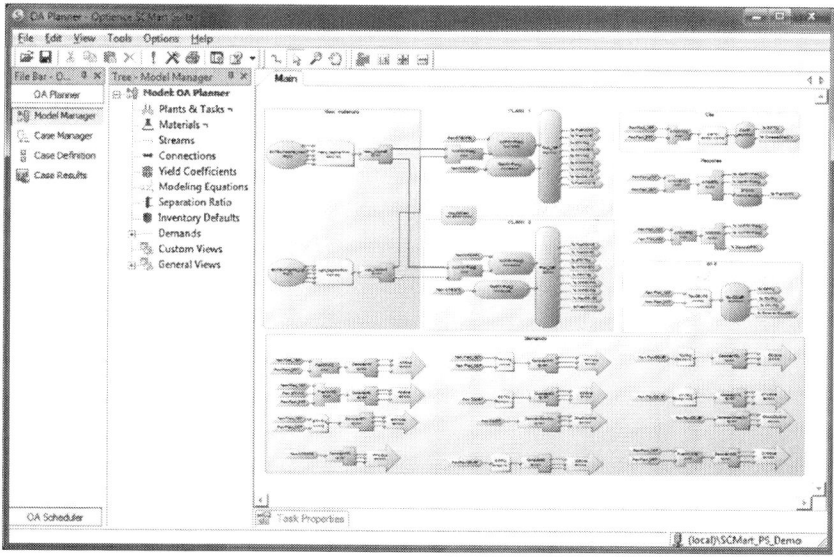

Figure 7.2: OA Planner process overview.

Another successful implementation was deployed at Copesul (today Braskem) in Brazil, which has two ethylene plants, two aromatics units and a C4s unit. The use of NLP and MINLP technology led to important benefits through the improvement of the operating conditions and the naphtha selection. A similar implementation was performed at another Braskem site in Salvador, BA. An NLP-based model was used to optimize the planning and scheduling of two ethylene plants, two aromatics and a reformer unit. The model also included the formulation of automotive gasoline, one C5s unit and two C4s units and consisted of more than 5000 variables per period. It allowed the company to make important market decisions by optimizing the olefins and reformer loads. Optience has implemented similar applications for other ethylene producers in Thailand and Indonesia.

Production Optimization in a Pulp and Paper Plant

One of the main challenges in the planning and scheduling of paper production is the simultaneous solution of the cutting stock problem. This should optimally match the customer order widths to the paper machine width with the aim of minimizing the trim loss, or in other words maximizing the productivity. Another concrete need stems from integrated paper mills, where one plant produces pulp and another one consumes the pulp for paper making. Here, not only the pulp mass balance must be synchronized, but there is also great potential for integrated energy management, since a chemical pulp plant produces typically more energy than a paper machine is using, and the production plan may also need to consider purchasing and selling electricity (Säynevirta & Luotojärvi, 2004). Some enterprises have centralized their energy planning across a number of plants (see Fig. 7.3).

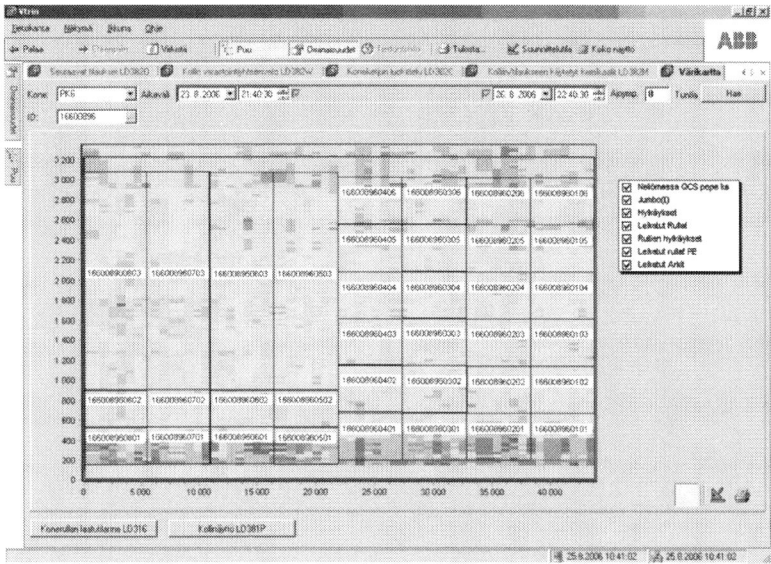

Figure 7.3: Quality view of a jumbo-reel with the cutting patterns and respective customer orders.

ABB CPM P&P Suite offers a solution that is able to manage most planning tasks (order planning, trimming, scheduling, quality, delivery planning, etc.) At one of the major paper companies estimated savings were 10–20 million EUR/year, mainly owing to a holistic planning and scheduling optimization approach, corresponding to a 2% improvement of the overall efficiency of the plant. The method automated the whole planning chain and replaced partly manual steps.

Crude-Oil Blend Scheduling Optimization

In the field of crude-oil blend scheduling, optimization can have a huge impact due to the large production volumes. Due to the mixing and splitting processes, nonlinear blending models, storage management, and pipeline availability, detailed modeling is required for the scheduling of a refinery's crude oil feedstocks from receipt to charging of the pipestills. Furthermore, the problem is highly complex because of the numerous crude oils that are finally processed into different products. Kelly and Mann (2003) discuss the economic and operational benefits associated with better crude oil blend scheduling.

Crude oils are often planned, purchased, procured and have a delivery schedule set long before they arrive at the refinery. Nevertheless, many practical aspects on off-loading, storing, blending and charging to meet pipestill feed quantity and quality specifications need to be based on current information, i.e. short-term requirements. The main potential benefits of optimized crude-oil blendshop scheduling comprise:

- Reduction of variability of quantity and quality targets
- Ability to generate several feasible and consistent schedules
- Better ability to react and adapt to spot-market opportunities
- Reduced pipeline penalties due to changes in sequence or timing
- Reduced working capital

Kelly and Mann (2003) report that optimization methods lead to approximate savings of 2.85 MUSD/year. This result could not be achieved manually due to the many couplings between sub-problems. Alone the problem logistics contains multiple aspects where main focus lies on quantity and logic variables. The quality subproblem is solved subsequently by fixing the values of the logic variables. This decomposition approach is well motivated by the large problem size, need to achieve a feasible solution to the logistics sub-problem, and the fact that if the logistics subproblem can be solved to global optimality then a respective feasible quality problem is the global optimum for the overall problem. For further details we refer to Kelly and Mann (2003) and the cited references.

Scheduling of Drumming Facility at Dow

The Dow Chemical Company has exploited the RTN approach to production scheduling for many years (Wassick, 2009 and Wassick and Ferrio, 2011). One successful application provides the schedule for a drumming facility that serves 15 different businesses shipping nearly 100 different products in total. This facility operates two automated drum fillers and one semi-automated tote and drum filler. Product to be drummed is drawn from storage tanks and railcars. A warehouse with limited capacity is used to store drums prior to shipment. Scheduling the drumming operation is very complex. The scheduler must take into account orders that include multiple products loaded into both drums and totes, coordination of rail car spotting, avoiding partial unloading of rail cars, warehouse capacity, due dates for orders and the availability of equipment due to shift schedules.

The discrete time RTN formulation is used to model the drumming operation. The key aspects of the scheduling problem, shared resources, equipment outages, inventory limits, due dates and complex tasks, are all easily modeled. For this particular application the RTN is dynamically built from demand data by postulating tasks that fill specific orders. This approach is not optimal in the purest sense. However it creates an RTN that can be solved in a reasonable amount of time and it produces a schedule that meets the expectations of the scheduler. Some of the tasks modeled in the RTN include drumming from a storage tank, drumming from a rail car, rail car spotting, multi-order drumming tasks, and transitions between products on a drumming line.

The tool is built as a custom Excel interface that extracts order information from SAP, builds the tasks using VBA macros, launches the RTN model on a remote server and displays the resulting schedule in a manner that allows the scheduler to make final adjustments before publishing the schedule. The use of the RTN based scheduling solution has resulted in a 75% reduction in scheduling related shutdowns of the warehouse with accompanying improved customer service levels and reduced use of third party drumming providers.

Scheduling in an Integrated Chemical Complex

The enterprise-wide optimization challenge at The Dow Chemical Company is discussed in Wassick (2009). The paper shows how an integrated chemical production site is composed of subsystems, called envelopes, which involve the production of a basic chemical and its downstream derivatives or comprise a network of units that provide an important utility. A key utility envelope is the waste treatment envelope, where an important operational challenge is the scheduling of liquid waste treatment. The system is composed of waste generating production plants, storage tanks, a complex network of piping and headers, waste treatment units, and loading stations for rail tank cars and tank trucks. Thus this is a network process, the main objective of which is to temporarily store and then process liquid waste generated by production plants. The goal is to never allow the waste treatment to become a constraint on production while operating the system within environmental constraints at lowest cost.

The waste treatment scheduling problem was modeled as a resource task network (RTN) MILP model. The objective function was to optimize operating cost and revenues through changes in the material resource levels. During the first year of operation the solutions obtained by the RTN model were compared against those obtained by the normal work process of the production planner, so that confidence on the model was gained. Fig. 7.4 shows that the optimization-based approach led to schedules with significantly lower cost than the solutions provided by the incumbent scheduling process. During the March–April time frame the model was able to produce a schedule that avoided sending waste off site for processing, the most costly treatment option.

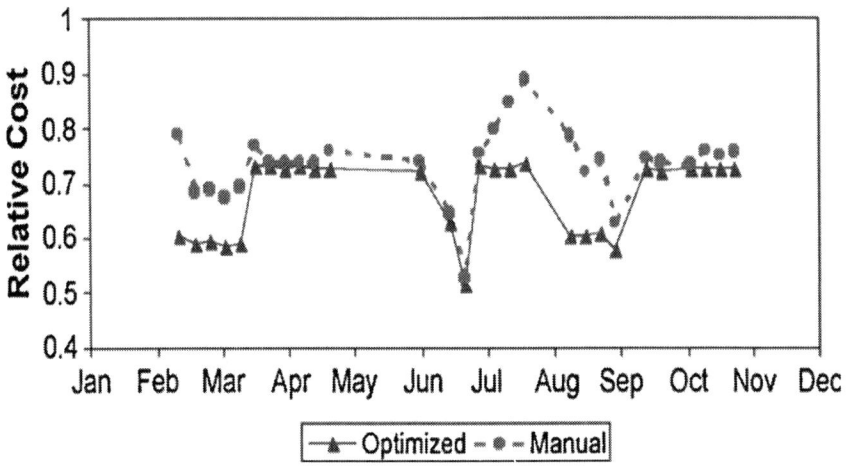

Figure 7.4: Comparison of an optimal schedule vs. a manual schedule.

The study also revealed that the planner favors transferring material directly from a storage tank to a treatment unit and tends to maintain the same destination day after day. All treatment units are operated at high load which offers the smallest marginal cost for finishing inputs. The optimal schedule completely rerouted the material flow and reduced the load of three treatment units. Tank-to-tank transfers occurred less often and the schedule was better than the planner's schedule because the cost of treatment chemicals used could be greatly reduced. This example shows the power of optimization and how it may identify better alternatives that conflict with intuitively best production strategies.

Medium-Term Scheduling of Large-Scale Chemical Plants

Floudas and co-workers have successfully tackled medium-term problems from large-scale batch and continuous plants using decomposition algorithms featuring a rolling-horizon scheme (Dimitriadis et al., 1997). Decisions are kept at manageable level, by selecting at a higher level the products to be considered on a particular subhorizon, which was done accounting for demand distribution, unit utilization and the limits on the complexity of the resulting scheduling problem (Lin, Floudas, Modi, & Juhasz, 2002). The lower level MILP scheduling models were based on the unit-specific continuous-time formulation ofIerapetritou and Floudas, 1998a and Ierapetritou and Floudas, 1998b, with the industrial processes being described as State-Task Networks.

The case study from ATOFINA Chemicals featured 35 products for a one-month production, made from a three-stage recipe involving the sharing of 10 equipment units (Lin et al., 2002). The batch plant from BASF involved the recipes of hundreds of products and over 80 pieces of equipment, with the case-study dealing with 67 products for a time horizon of 19 days (Janak et al., 2006a and Janak et al., 2006b). Another chemical plant from BASF comprised up to 100 units, being 1/3 processing and 2/3 storage units, for producing 100 different products over a one month time horizon (Shaik et al., 2009).

Graphical user interfaces were developed to integrate the various components required to apply the optimization framework. Software for the ATOFINA problem was developed in Visual Basic and has the following main features: (i) link to an Access database for storing all the information and changing it throughout the iterative procedure; (ii) link to the CPLEX solver for solving the MILP scheduling problems; (iii) solution output in the form of a Gantt chart. Software for the BASF problems is used on a daily basis to improve planning and scheduling by ensuring an efficient utilization of the main units. It features implementation in GAMS/CPLEX with new interfacing technology and integration approaches to connect to SAP-APO.

Summary

A number of successful applications of optimized scheduling have been briefly discussed. It is important to highlight the fact that an optimal schedule and production plan can only bring true benefits when the production is aligned to it. Often only partially following an optimized schedule is not sufficient and may even lead to worse solutions than before. Therefore, it is crucial that the planning system is able to capture the major decisions in a correct way and the user expectations and needs are fully aligned with the system and its user interfaces.

EMERGING INDUSTRIAL PROBLEMS

This paper has emphasized the industrial implementation and application of scheduling methodologies at the plant floor. One problem is that industrial requirements are periodically changing due to e.g. pressure toward further cost-savings, regulatory aspects, appearing new technologies and standards, as well as changing competitive situation. In the following some emerging challenges are discussed, which also have a significant impact on production planning and scheduling optimization:

- How to deal with energy? Production should be more energy efficient, flexible, and agile and able to adapt to situations where energy supply is restricted. This is a major focus e.g. in the research on industrial demand-side management. So far major contributions in this area have been on the power producer and grid management side (unit commitment) but the focus is increasingly expanding into production processes e.g. in process industries. The possible roles and importance of scheduling optimization are still only partly clarified.

- Directly related to above, electricity pricing and its immediate impact on production scheduling and potential new opportunities have not been studied sufficiently. With an increasing amount of renewable (partly unpredictable) energy production, prices will be facing higher fluctuations in the future. The ability to consider true energy costs can enable a significant savings potential, especially in energy-intensive industries.

- Raw-material pricing and availability has been more volatile than ever in the last decade. For instance crude oil prices have undergone severe changes affecting almost all industries. Industries, such as steel and electronics, have been faced with uncertain price developments (nickel and lithium). Due to increased biofuels production, scarcity of certain raw-materials is affecting the food as well as the pulp and paper industry. The ability to consider also the feedstock prices as a natural part of planning and scheduling may result into situations where the most economical strategy is to temporarily suspend part or the whole production.

- Regulations on emissions restrict the production especially in the developed countries. Various emissions (CO_2, NOX, SO_2) may be limited by governmental or local regulations, pricing or downstream processes. This all adds a novel component into the planning that is difficult to manage without sufficient tools.

- Globalization has increased the transportation costs and multi-site planning is becoming more common. If several locations across the supply chain need to be simultaneously considered, the size of resulting scheduling problems also increases significantly. This calls for more efficient algorithms.

- How to efficiently use the existing and ever increasing data and knowledge to support the scheduling process? Most companies have their best practices which should be reflected in the decision making to speed up the development of new approaches and avoid potential conflicts. Is this only a big-data issue or does it require a major paradigm change is an open question.

- Due to rapid market changes processes need to be more flexible and the design of a process may need to be adapted on-the-fly to enable the production of various product portfolios. Planning and scheduling solutions can support this through more flexible production systems.

- Measuring the "goodness" of a solution is vital, especially in cases where a number of alternative production plans need to be compared and possibly manually selected. The challenge is how to represent costs and benefits in a generic way and balance the problem objectives vs. some user defined KPIs?

Many of these challenges can already today be handled through existing scheduling optimization approaches in some form, at least in small scale. For instance, discrete-time approaches are by nature well suited for tackling resource constraints. The main complexity, however, comes from combining the challenges. For instance the exactness of discrete-time formulations is sometimes not sufficient–or the size of the planning horizon is too short as only a certain number of time grid points can be handled efficiently. Methodologies combining optimization approaches with heuristics or simulation and algorithms built on solving subsequent problems improving iteratively provide a vast field of research. The most important issue is to collaborate across "traditional" borders and step-wise advance toward solutions that can respond to most industrial needs. Here it is crucial that:

- Industry offers enough real-life problems to the research community and helps evaluating the results in order to guideline the research to look for the right KPIs.

- Academia values the importance of practical applicability of methods as to ensure fast solution of large and complicated problems, not necessarily to optimality.

- Various research fields and communities join in projects to merge methodologies where certain strengths can lend themselves to collaborative solutions.

Following the three above aspects can also enable quicker resolution of most challenges.

CONCLUSIONS AND FUTURE DIRECTIONS

This paper has given an overview of the scheduling area and focused on the industrial applicability of scheduling models and methods. Since no single approach exists that fits to all problems, one evident challenge is to choose the best theoretical method or approach for tackling the real problem at hand. Practically all processing features can be theoretically modeled and there are methods to address problems in all production environments. Thus, some of the main challenges are the computational complexity and cost of implementation, and here selecting the right modeling and solution method is of key. However, the implementation of these also poses important challenges when aiming at ease-of-use and lowering the engineering costs necessary to adapt the method to a specific problem. Here, standards (e.g. ISA-95) can be helpful and the use of systematic approaches for model reformulation can dramatically improve the re-usability. Ideally, we would like to be able build models such that they can be simply configured to new problems without touching the algorithmic core. However, scalability remains a problem as the combinatorial complexity of large problem instances does not disappear. Thus it is important to understand the limitations of optimization methods and algorithms.

Nevertheless, the challenges are not merely technical but more multifaceted. For instance, the number of optimization specialists is often limited. In fact, most engineers and operators lack sufficient exposure to optimization so that the acceptance of its potential may be limited by both at the management level as well as at the plant level. As implementation projects are wide and information technology plays an important role, one question is what the optimal mix of expertise, e.g. process engineers, OR-experts and programmers should be. A true production management system should comprise scheduling as a natural part of a well-orchestrated environment, which raises many challenges on how to interface and compromise between various – potentially competing – optimization functions.

Although we are not quite there yet, we do hope that we have managed to convince the reader that there is high potential, even more a need to further develop and apply optimization solutions to meet the upcoming challenges. Also, many promising steps have been taken into the right direction, which can be seen of the reported success stories. The main question is how to efficiently bring the various disciplines together to raise the state-of-the-art to the next level, where implementation of scheduling methodologies into real industrial problems is common practice.

ACKNOWLEDGMENTS

Pedro Castro acknowledges financial support from FEDER (Programa Operacional Factores de Competitividade – COMPETE) and Fundação para a Ciência e Tecnologia through project FCOMP-01-0124-FEDER-020764. Christos Maravelias acknowledges support from the National Science Foundation under Grant CBET-1066206. Ignacio Grossmann acknowledges support from the National Science Foundationunder Grant CBET-1159443. We also want to acknowledge Dr. Arturo Cervantes for providing us valuable information on Optience planning and scheduling tools.

REFERENCES

1.   Abdeddaim, Y., & Maler, O. (2001). Job-shop scheduling using timed automata. Computer aided verification; LNCS (Vol. 2102) Berlin: Springer.

2.   Abhishek, K., Leyffer, S., & Linderoth, J. T. (2010). FilMINT: An outer approximationbased solver for convex mixed-integer nonlinear programs. INFORMS Journal on Computing, 22, 555–567.

3.   Achterberg, T. (2008). SCIP: Solving constraint integer programs. Mathematical Programming Computation, 1, 1–41

4.   Ahuja, R. K., Ergun, Ö., Orlin, J. B., & Punnen, A. P. (2002). A survey of very large scale neighborhood search techniques. Discrete Applied Mathematics, 123, 75–102.

5.   Alur, R., & Dill, D. L.(1992). The theory oftimed automata. Real-time: Theory in practice; LNCS (Vol. 600) Berlin: Springer.

6.   Anderson, E. J., & Nyirendra, J. C. (1990). Two new rules to minimize tardiness in a jobshop. International Journal of Production Research, 28, 2277–2292.

7.   ANSI/ISA-95.00.03-2005.(2005). Enterprise-control system integration. Part 3:Activity models of manufacturing operations management. , ISBN 1-55617-955-3

8.   Aron, I., Hooker, J. N., & Yunes, T. H. (2004). SIMPL: A system for integrating optimization techniques. In J.-C. Régin, & M. Rueher (Eds.), Conference on integration of AI and OR techniques in constraint programming for combinatorial optimization problems (CPAIOR 2004), LNCS 3011 (pp. 21–36). Springer.

9.   Atamturk, A., & Savelsbergh, M. W. P. (2005). Integer-programming software systems. Annals of Operations Research, 140(1), 67–124.

10. Baker, K. R. (1974). Introduction to sequencing and scheduling. New York: J. Wiley. Balas, E., Lenstra, J. K., & Vazacopoulos, A. (1995). The one-machine problem with delayed precedence constraints and its use in job shop scheduling. Management Science, 41, 94–109.

11. Balasubramanian, J., & Grossmann, I. E. (2002). A novel branch and bound algorithm for scheduling flowshop plants with uncertain processing times. Computers & Chemical Engineering, 26(1), 41–57.

12. Balasubramanian, J., & Grossmann, I. E. (2003). Scheduling optimization under uncertainty – An alternative approach. Computers & Chemical Engineering, 27(4), 469–490.

13. Bassett, M. H., Pekny, J. F., & Reklaitis, G. V. (1996). Decomposition techniques for the solution of large-scale scheduling problems. AIChE Journal, 42, 3373–3387.

14. Beldiceanu, N., Carlsson, M., & Rampon, J.-X. (2010). Global Constraint Catalog, SICS Technical Report T2010:07. Stockholm: Swedish Institute of Computer Science. Online version at. http://www.emn.fr/z-info/sdemasse/gccat/

15. Belotti, P., Lee, J., Liberti, L., Margot, F., & Wächter, A. (2009). Branching and bounds tightening techniques for non-convex MINLP. Optimization Methods and Software, 24(4), 597–634.

16. Ben-Tal, A., El Ghaoui, L., & Nemirovski, A. (2009). Robust optimization. Princeton, NJ: Princeton University Press.

17. Bhushan, S., & Karimi, I. A. (2003). An MILP approach to automated wet-etch station scheduling. Industrial & Engineering Chemistry Research, 42, 1391–1399.

18. Birge, J. R., & Louveaux, F. (1997). Introduction to stochastic programming. New York: Springer.

19. Bixby, R., & Rothberg, E. (2007). Progress in computational mixed integer programming: A look back from the other side of the tipping point. Annals of Operations Research, 149(1), 37–41.

20. Blum, C., Puchinger, J., Raidl, G., & Roli, A. (2011). Hybrid metaheuristics. In P. van Hentenryck, & M. Milano (Eds.), Hybrid optimization: The ten years of CPAIOR (pp. 305–336). Cambridge, MA: Springer.

21.   Bockmayr, A., & Kasper, T. (1998). Branch-and-infer: A unifying framework for integer and finite domain constraint programming. INFORMS Journal on Computing, 10, 287–300.

22.   Bockmayr, A., & Kasper, T. (2004). Branch-and-infer. In M. Milano (Ed.), Constraint and integer programming: Toward a unified methodology (pp. 59–88). Springer.

23.   Bonami, P., Biegler, L. T., Conn, A. R., Cornuejols, G., Grossmann, I. E., Laird, C. D., et al. (2008). An algorithmic framework for convex mixed integer nonlinear programs. Discrete Optimization, 5, 186–204.

24.   Bongers, P. M. M., & Bakker, B. H. (2006). Application of multi-stage scheduling. In ESCAPE 16 proceedings.

25.   Bongers, P. M. M., & Bakker, B. H.(2008).Validation of an Ice cream factory operations model. In ESCAPE 18 proceedings.

26.   Brooke, A., Kendrick, D., & Meeraus, A. (1988). GAMS: A User's Guide. San Francisco, CA: The Scientific Press.

27.   Burkard, R. E., & Hatzl, J. (2005). Review: Extensions and computational comparison of MILP formulations for scheduling of batch processes. Computers & Chemical Engineering, 29(8), 1752–1769.

28.   Cafaro, D. C., & Cerdá, J.(2004). Optimal scheduling of multiproduct pipeline systems using a non-discrete MILP formulation. Computers & Chemical Engineering, 28, 2053.

29.   Calfa, B. A., Agarwal, A., Grossmann, I. E., & Wassick, J. M. (2013). Hybrid bilevel-lagrangean decomposition scheme for the integration of planning and scheduling of a network of batch plants. Industrial & Engineering Chemistry Research, 52, 2152–2167.

30.   Capón-Garcia, E., Ferrer-Nadal, S., Graells, M., & Puigjaner, L.(2009).An extended formulation for the flexible short-term scheduling of multiproduct semicontinuous plants. Industrial & Engineering Chemistry Research, 48, 2009–2019.

31.   Carlier, J. (1982). The one-machine sequencing problem. European Journal of Operational Research, 11, 42–47.

32.   Castro, P. M. (2010). Optimal scheduling of pipeline systems with a resourcetask network continuous-time formulation. Industrial & Engineering Chemistry Research, 49, 11491–11505.

33. Castro, P., Barbosa-Póvoa, A. P. F. D., & Matos, H. (2001). An improved RTN continuous-time formulation for the short-term scheduling of multipurpose batch plants. Industrial & Engineering Chemistry Research, 40, 2059–2068.

34. Castro, P. M., Barbosa-Póvoa, A. P., & Matos, H. A.(2003). Optimal periodic scheduling of batch plants using RTN-based discrete and continuous-time formulations: A case study approach. Industrial & Engineering Chemistry Research, 42, 3346.

35. Castro, P. M., Barbosa-Póvoa, A. P., Matos, H. A., & Novais, A. Q. (2004). Simple continuous-time formulation for short-term scheduling of batch and continuous processes. Industrial & Engineering Chemistry Research, 43, 105–118.

36. Castro, P. M., Erdirik-Dogan, M., & Grossmann, I. E. (2008). Simultaneous batching and scheduling of single stage batch plants with parallel units. AIChE Journal, 54, 183–193.

37. Castro, P. M., & Grossmann, I. E. (2005). New continuous-time MILP model for the short-term scheduling of multistage batch plants. Industrial & Engineering Chemistry Research, 44, 9175–9190.

38. Castro, P. M., Grossmann, I. E., & Novais, A. Q. (2006). Two new continuous-time models for the scheduling of multistage batch plants with sequence dependent changeovers. Industrial & Engineering Chemistry Research, 45, 6210–6226.

39. Castro, P. M., & Grossmann, I. E. (2012). Generalized disjunctive programming as a systematic modeling framework to derive scheduling formulations. Industrial & Engineering Chemistry Research, 51, 5781–5792.

40. Castro, P. M., Harjunkoski, I., & Grossmann, I. E. (2009). New continuous-time scheduling formulation for continuous plants under variable electricity cost. Industrial & Engineering Chemistry Research, 48, 6701–6714.

41. Castro, P. M., Harjunkoski, I., & Grossmann, I. E. (2011). Greedy algorithm for scheduling batch plants with sequence-dependent changeovers. AIChE Journal, 57, 373–387.

42. Castro, P., Westerlund, J., & Forssell, S. (2009). Scheduling of a continuous plant with recycling of byproducts: A case study from a tissue paper mill. Computers & Chemical Engineering, 33, 347–358.

43. Castro, P. M., Zeballos, L. J., & Méndez, C. A. (2012). Hybrid time slots sequencing model for a class of scheduling problems. AIChE Journal, 58(3), 789–800.

44. Chen, P., Papageorgiou, L. G., & Pinto, J. M. (2008). Medium-term planning of singlestage single-unit multiproduct plants using a hybrid discrete/continuous-time MILP model. Industrial & Engineering Chemistry Research, 47, 1925–1934.

45. Colombani, Y., & Heipcke, S. (2002). Mosel: An extensible environment for modeling and programming solutions. In N. Jussien, & F. Laburthe (Eds.), CPAIOR proceedings Springer, (pp. 277–290).

46. Cornuejols, G. (2008). Valid inequalities for mixed integer linear programs. Mathematical Programming, 112(1), 3–44.

47. Cui, J., & Engell, S. (2010). Medium-term planning of a multiproduct batch plant under evolving multi-period multi-uncertainty by means of a moving horizon strategy. Computers & Chemical Engineering, 34, 598–619.

48. Dimitriadis, A. D., Shah, N., & Pantelides, C. C. (1997). RTN-based rolling horizon algorithms for medium term scheduling of multipurpose plants. Computers & Chemical Engineering, 21, S1061.

49. Duarte, B., Santos, L., & Mariano, J. (2009). Optimal sizing, scheduling and shift policy of the grinding section of a ceramic tile plant. Computers Operations. Research, 36, 1825–1834.

50. Easton, K., Nemhauser, G., & Trick, M.(2004). CP based branch and price. In M. Milano (Ed.), Constraint and integer programming: Toward a unified methodology (pp. 207–231). Kluwer.

51. Engell, S., & Harjunkoski, I. (2012). Optimal operation: Scheduling, advanced control and their integration. Computers & Chemical Engineering, 47, 121–133.

52. Erdirik-Dogan, M., & Grossmann, I. E. (2006). A decomposition method for the simultaneous planning and scheduling of single-stage continuous multiproduct plants. Industrial & Engineering Chemistry Research, 45, 299–315.

53. Fehnker, A. (1999). Scheduling a steel plant with timed automata. In Proceedings of the 6th international conference on real-time computing systems and applications (pp. 280–286).

54. Ferrer-Nadal, S., Capón-Garcia, E., Méndez, C. A., & Puigjaner, L. (2008). Material transfer operations in batch scheduling. A critical modeling issue. Industrial & Engineering Chemistry Research, 47, 7721–7732.

55. Ferris, M., Maravelias, C. T., & Sundaramoorthy, A. (2009). Simultaneous batching and scheduling using dynamic decomposition on a grid. INFORMS Journal on Computing, 21(3), 398–410.

56. Fischetti, M., Glover, F., & Lodi, A. (2005). The feasibility pump. Mathematical Programming, 104, 91–104.

57. Fischetti, M., & Lodi, A. (2008). Repairing MIP infeasibility through local branching. Computers and Operations Research, 35, 1436–1445.

58. Flores-Tlacuahuac, A., & Grossmann, I. E.(2006). Simultaneous cyclic scheduling and control of a multiproduct CSTR. Industrial & Engineering Chemistry Research, 45, 6698–6712.

59. Floudas, C. A., & Lin, X. (2004). Continuous-time versus discrete-time approaches for scheduling of chemical processes: A review. Computers & Chemical Engineering, 28, 2109–2129.

60. Frost, & Sullivan.(2010).World Manufacturing Execution Systems (MES) Market, N668- 10, October 2010.

61. Georgiadis, M. C., Levis, A. A., Tsiakis, P., Sanidiotis, L., Pantelides, C. C., & Papageorgiou, L. G.(2005). Optimisation-based scheduling:Adiscretemanufacturing case study. Computers & Industrial Engineering, 49(1), 118–145.

62. Georgiadis, M. C., Papageorgiou, L. G., & Macchietto, S. (2000). Optimal cleaning policies in heat exchanger networks under rapid fouling. Industrial & Engineering Chemistry Research, 39(2), 441–454.

63. Gimenez, D. M., Henning, G. P., & Maravelias, C. T. (2009a). A novel networkbased continuous-time representation for process scheduling: Part I. Main concepts and mathematical formulation. Computers & Chemical Engineering, 33(9), 1511–1528.

64. Gimenez, D. M., Henning, G. P., & Maravelias, C. T. (2009b). A novel network-based continuous-time representation for process scheduling: Part II. General framework. Computers & Chemical Engineering, 33(10), 1644–1660.

65. Glismann, K., & Gruhn, G. (2001). Short-term scheduling and recipe optimization of blending processes. Computers & Chemical Engineering, 25, 627–634.

66. Grossmann, I. E. (2002). Review of nonlinear mixed-integer and disjunctive programming

67. techniques. Optimization and Engineering, 3, 227–252.

68. Harjunkoski, I., & Grossmann, I. (2002). Decomposition techniques for multistage scheduling problems using mixed-integer and constraint programming methods. Computers & Chemical Engineering, 26, 1533–1552.

69. Haupt, R. (1989). A survey of priority rule-based scheduling. OR Spektrum, 11, 3–16.

70. Hollender, M. (Ed.). (2009). Collaborative process automation systems. , ISBN 978- 1936007103. Research Triangle Park, NC: ISA.

71. Holthaus, O., & Rajendran, C. (2000). Efficient jobshop dispatching rules: Further developments. Production Planning and Control, 11, 171–178.

72. Honkomp, S. J., Mockus, L., & Reklaitis, G. V. (1999). A Framework for schedule evaluation with processing uncertainty. Computers & Chemical Engineering, 23(4–5), 595–609.

73. Hooker, J. N. (2006). An integrated method for planning and scheduling to minimize tardiness. Constraints, 11, 139–157.

74. Hooker, J. N.(2007). Planning and scheduling by logic-based Benders decomposition. Operations Research, 55, 588–602.

75. Hooker, J. N. (2012). Integrated methods for optimization (2nd ed.). Springer.

76. Hooker, J. N., & Osorio, M. A. (1999). Mixed logical/linear programming. Discrete Applied Mathematics, 96–97, 395–442.

77. Hooker, J. N., & Ottosson, G. (2003). Logic-based Benders decomposition. Mathematical Programming, 96, 33–60.

78. Hooker, J. N., & Yan, H. (1995). Logic circuit verification by Benders decomposition. In V. Saraswat, & P. Van Hentenryck (Eds.), Principles and practice of constraint programming: The Newport papers (pp. 267–288). Cambridge, MA: MIT Press.

79. Horst, R., & Tuy, H. (1996). Global optimization deterministic approaches (3rd ed.). Berlin: Springer-Verlag.

80. Ierapetritou, M. G., & Floudas, C. A. (1998a). Effective continuous-time formulation for short-term scheduling: 1. Multipurpose batch processes. Industrial and Engineering Chemistry Research, 37, 4341–4359.

81. Ierapetritou, M. G., & Floudas, C. A. (1998b). Effective continuous-time formulation for short-term scheduling: 2. Continuous and semi-continuous processes. Industrial & Engineering Chemistry Research, 37, 4360.

82. Janak, S. L., & Floudas, C. A. (2008). Improving unit-specific event based continuoustime approaches for batch processes: Integrality gap and task splitting. Computers & Chemical Engineering, 32(4–5), 913–955.

83. Janak, S. L., Floudas, C. A., Kallrath, J., & Vormbrock, N. (2006a). Production scheduling of a large-scale industrial batch plant. I. Short-term and medium-term scheduling. Industrial & Engineering Chemistry Research, 45, 8234.

84. Janak, S. L., Floudas, C. A., Kallrath, J., & Vormbrock, N. (2006b). Production scheduling of a large-scale industrial batch plant. II. Reactive scheduling. Industrial & Engineering Chemistry Research, 45, 8253.

85. Janak, S. L., Lin, X., & Floudas, C. A. (2007). A new robust optimization approach for scheduling under uncertainty: II. Uncertainty with known probability distribution. Computers & Chemical Engineering, 31, 171–195.

86. Jain, V., & Grossmann, I. E. (1998). Cyclic scheduling and maintenance of parallel process units with decaying performance. AIChE Journal, 44, 1623–1636.

87. Jain, V., & Grossmann, I. E. (2001). Algorithms for hybrid MILP/CP models for a class of optimization problems. INFORMS Journal on Computing, 13(4), 258–276.

88. Johnson, E. L., Nemhauser, G. L., & Savelsbergh, M. W. P. (2000). Progress in linear programming-based algorithms for integer programming: An exposition. INFORMS Journal on Computing, 12(1), 2–23.

89. Junker, U., Karish, S. E., Kohl, N., Vaaben, B., Fahle, T., & Sellmann, M. (1999). A framework for constraint programming based column generation. In J. Jaffar (Ed.), Principles and practice

of constraint programming (CP 1999), LNCS [1713] (pp. 261–275). Springer.

90. Kallrath, J. (2002). Planning and scheduling in the process industry. OR Spektrum, 24,

91. 219–250.

92. Kelly, J. D. (2006). Logistics: The missing link in blend scheduling optimization. Hydrocarbon Processing, 85(6), 45–51.

93. Kelly, J. D., & Mann, J. L. (2003). Crude oil blend scheduling optimization: An application with multimillion dollar benefits. Hydrocarbon Processing, 82, 47–53.

94. Kelly, J. D., & Zyngier, D. (2008). Hierarchical decomposition heuristic for scheduling: Coordinated reasoning for decentralized and distributed decision-making problems. Computers & Chemical Engineering, 32(11), 2684–2705.

95. Kolodziej, S. P., Grossmann, I. E., Furman, K. C., & Sawaya, N. W. (2013). A discretization-based approach for the optimization of the multiperiod blend scheduling problem. Computers & Chemical Engineering, 53, 122–142.

96. Kondili, E., Pantelides, C. C., & Sargent, W. H. (1993). A general algorithm for shortterm scheduling of batch operations – I. MILP formulation. Computers & Chemical Engineering, 2, 211–227.

97. Kopanos, G. M., Capon-Garcia, E., Espuna, A., & Puigjaner, L. (2008). Costs for rescheduling actions: A critical issue for reducing the gap between scheduling theory and practice. Industrial Engineering Chemistry Research, 47(22), 8785–8795.

98. Kopanos, G. M., Laínez, J. M., & Puigjaner, L. (2009). An efficient mixed-integer linear programming scheduling framework for addressing sequence-dependent setup issues in batch plants. Industrial & Engineering Chemistry Research, 48, 6346–6357.

99. Kopanos, G. M., Méndez, C. A., & Puigjaner, L. (2010). MIP-based decomposition strategies for large-scale scheduling problems in multiproduct multistage batch plants: A benchmark scheduling problem of the pharmaceutical industry. European Journal of Operation Research, 207, 644–655.

100. Kopanos, G. M., Puigjaner, L., & Georgiadis, M. C. (2012). Efficient mathematical frameworks for detailed production scheduling in food processing industries. Computers & Chemical Engineering, 42, 206–216.

101. Kopanos, G. M., Puigjaner, L., & Maravelias, C. T. (2011). Production planning and scheduling of parallel continuous processes with product families. Industrial & Engineering Chemistry Research, 50, 1369–1378.

102. Land, A. H., & Doig, A. G. (1960). An automatic method of solving discrete programming problems. Econometrica, 28(3), 497–520.

103. Larsen, K. G., Behrmann, G., Brinksma, E., Fehnker, A., Hune, T., Pettersson, P., et al. (2001). As cheap as possible: Efficient cost-optimal reachability for priced timed automata. In Computer aided verification, LNCS 2102. Berlin: Springer.

104. Lee, H., Pinto, J. M., Grossmann, I. E., & Park, S. (1996). MILP model for refinery short term scheduling of crude oil unloading with inventory management. I&EC Research, 35, 1630–1641.

105. Leyffer, S. (2001). Integrating SQP and branch and bound for mixed integer nonlinear programming. Computational Optimization and Applications, 18, 295.

106. Li, J., & Floudas, C. A. (2010). Optimal event point determination for shortterm scheduling of multipurpose batch plants via unit-specific event-based continuous-time approaches. Industrial & Engineering Chemistry Research, 49, 7446–7469.

107. Li, Z., & Ierapetritou, M. (2008a). Process scheduling under uncertainty: Review and challenges. Computers & Chemical Engineering, 32, 715–727.

108. Li, Z. K., & Ierapetritou, M. G. (2008b). Reactive scheduling using parametric programming. AIChE Journal, 54(10), 2610–2623.

109. Lin, X., Floudas, C. A., Modi, S., & Juhasz, N. M. (2002). Continuous-time optimization approach for medium-range production scheduling of a multiproduct batch plant. Industrial & Engineering Chemistry Research, 41, 3884.

110. Lin, X., Janak, S. L., & Floudas, C. A. (2004). A new robust optimization approach for scheduling under uncertainty: I. Bounded uncertainty. Computers & Chemical Engineering, 28, 1069–1108.

111. Lindo Systems Inc. (2010). LindoGLOBAL Solver.

112. Liu, Y., & Karimi, I. A. (2007). Scheduling multistage, multiproduct batch plants with nonidentical parallel units and unlimited

intermediate storage. Chemical Engineering Science, 62, 1549–1566.

113. Liu, Y., & Karimi, I. A. (2008). Scheduling multistage batch plants with parallel units and no interstage storage. Computers & Chemical Engineering, 32, 671–693.

114. Liu, S. S., Pinto, J. M., & Papageorgiou, L. G. (2008). A TSP-based MILP model for medium-term planning of single-stage continuous multiproduct plants. Industrial & Engineering Chemistry Research, 47, 7733–7743.

115. Malapert, A., Gueret, C., & Rousseau, L. M. (2012). A constraint programming approach for a batch processing problem with non-identical job sizes. European Journal of Operational Research, 221(3), 533–545.

116. Maravelias, C. T. (2006). A decomposition framework for the scheduling of single- and multi-stage processes. Computers & Chemical Engineering, 30, 407–420.

117. Maravelias, C. T. (2012). General framework and modeling approach classification for chemical production scheduling. AIChE Journal, 58(6), 1812–1828.

118. Maravelias, C. T., & Grossmann, I. E. (2003a). A new general continuous-time state task network formulation for short term, scheduling of multipurpose batch plants. I&EC Research, 42, 3056–3074.

119. Maravelias, C.T., & Grossmann, I. E.(2003b). A general continuous state task network formulation for short term scheduling of multipurpose batch plants with due dates. In A. Kraslawski, & I. Turunen (Eds.), Computer-aided chemical engineering (Vol. 14) (p. 215). Amsterdam, The Netherlands: Elsevier.

120. Maravelias, C. T., & Grossmann, I. E. (2004). A hybrid MILP/CP decomposition approach for the continuous time scheduling of multipurpose batch plants. Computers & Chemical Engineering, 28, 1921–1949.

121. Maravelias, C. T., & Sung, C. (2009). Integration of production planning and scheduling: Overview, challenges and opportunities. Computers&Chemical Engineering, 33(12), 1919–1930.

122. Méndez, C. A., & Cerdá, J. (2002). An MILP framework for short-term scheduling

123. of single-stage batch plants with limited discrete resources. In J. Grievink, &

124. J. van Schijndel (Eds.), Computer-aided chemical engineering (Vol. 10) (p. 721).

125. Amsterdam, The Netherlands: Elsevier.

126. Méndez, C. A., & Cerda, J. (2003a). An MILP continuous-time framework for shortterm scheduling of multipurpose batch processes under different operation strategies. Optimization and Engineering, 4, 7–22.

127. Méndez, C. A., & Cerda, J. (2003b). Dynamic scheduling in multiproduct batch plants. Computers & Chemical Engineering, 27(8–9), 1247–1259.

128. Méndez, C. A., Cerdá, J., Grossmann, I. E., Harjunkoski, I., & Fahl, M. (2006). Stateof-the-art review of optimization methods for short-term scheduling of batch processes. Computers & Chemical Engineering, 30, 913–946.

129. Méndez, C. A., Grossmann, I. E., Harjunkoski, I., & Kabore, P. (2006). A simultaneous optimization approach for off-line blending and scheduling of oil-refinery operations. Computers & Chemical Engineering, 30, 614–634.

130. Méndez, C. A., Henning, G. P., & Cerda, J. (2001). An MILP continuous time approach to short-term scheduling of resource constrained multi-stage flowshop batch facilities. Computers & Chemical Engineering, 25, 701–711.

131. Michel, L., See, A., & Van Hentenryck, P. (2009). Parallel and distributed local search in comet. Computers and Operations Research, 36, 2357–2375.

132. Misener, R., & Floudas, C. A. (2010). Global optimization of large-scale generalized pooling problems: Quadratically constrained MINLP models. Industrial & Engineering Chemistry Research, 49(11), 5424–5438.

133. Misener, R., & Floudas, C. (2013). GloMIQO: Global mixed-integer quadratic optimizer. Journal of Global Optimization, 57(1), 3–50.

134. Mishra, B. V., Mayer, E., Raisch, J., & Kienle, A. (2005). Short-term scheduling of batch processes. A comparative study of different approaches. Industrial & Engineering Chemistry Research, 44, 4022–4034.

135. Moser, M., & Engell, S. (1992). A survey of priority rules for FMS scheduling and their performance for the benchmark problem. In Proceedings of the 31st IEEE international conference on decision and control (pp. 392–397).

136. Moser, M., Herrmann, M., & Engell, S. (1992). Avoiding scheduling errors by partial simulation of the future. In Proceedings of the 31st IEEE international conference on decision and control Tucson, (pp. 411–412).

137. Mouret, S., Grossmann, I. E., & Pestiaux, P. (2009). A novel priority-slot based continuous-time formulation for crude-oil scheduling problems. Industrial & Engineering Chemistry Research, 48, 8515–8528.

138. Mouret, S., Grossmann, I. E., & Pestiaux, P. (2011). Time representations and mathematical models for process scheduling problems. Computers and Chemical Engineering, 35(6), 1038–1063.

139. Nemhauser, G. L., & Wolsey, L. A. (1988). Integer and combinatorial optimization. New York: Wiley. Novas, J. M., & Henning, G. P. (2010). Reactive scheduling framework based on domain knowledge and constraint programming. Computers & Chemical Engineering, 34(12), 2129–2148.

140. Novas, J. M., & Henning, G. P. (2012). A comprehensive constraint programming approach for the rolling horizon-based scheduling of automated wet-etch stations. Computers & Chemical Engineering, 42, 189–205.

141. Nystrom, R. H., Franke, R., Harjunkoski, I., & Kroll, A. (2005). Production campaign planning including grade transitionsequencing anddynamic optimization. Computers & Chemical Engineering, 29(10), 2163–2179.

142. Panek, S., Engell, S., Subbiah, S., & Stursberg, O. (2008). Scheduling of multi-product batch plants based upon timed automata models. Computers & Chemical Engineering, 32, 275–291.

143. Panek, S., Stursberg, O., & Engell, S. (2006). Efficient synthesis of production schedules by optimization of timed automata. Control Engineering Practice, 14, 1183–1197.

144. Pantelides, C. C.(1994). Unified frameworks for optimal process planning and scheduling. In Foundations of computer-aided process operations. New York: CACHE Publications.

145. Panwalkar, S. S., & Iskander, W. (1977). A survey of scheduling rules. Operations Research, 25(1), 45–61.

146. Papageorgiou, L. G., & Pantelides,C.C.(1996). Optimal campaignplanning scheduling of multipurpose batch semicontinuous plants. 2. A mathematical decomposition approach. Industrial & Engineering Chemistry Research, 35, 510–529.

147. Pekny, J. F., Miller, D. L., & McRaec, G. J. (1990). An exact parallel algorithm for scheduling when production costs depend on consecutive system states. Computers & Chemical Engineering, 14, 1009–1023.

148. Phanden,R.K.,Jain,A., & Verma,R.(2011).Integration of process planning and scheduling: A state-of-the-art review. International Journal of Computer Integrated Manufacturing, 24(6), 517–534.

149. Piana, S., & Engell, S. (2011). Hybrid evolutionary optimization of the operation of pipeless plants. Journal of Heuristics, 16, 311–336.

150. Pierreval, H., & Mebarki, N. (1997). Dynamic selection of dispatching rules for manufacturing system scheduling. International Journal of Production Research, 35, 1575–1591.

151. Pinedo, M., & Chao, X. (1999). Operations scheduling with applications in manufacturing and services. Boston: Irwin/McGraw-Hill., ISBN 0-07-289779-1.

152. Pinto, J., & Grossmann, I. E. (1994). Optimal cyclic scheduling of multistage multiproduct continuous plants. Computers & Chemical Engineering, 18, 797–816.

153. Pinto, J., & Grossmann, I. A. (1995). Continuous time mixed integer linear programming model for the short-term scheduling of multistage batch plants. Industrial & Engineering Chemistry Research, 34, 3037.

154. Pinto, J. M., & Grossmann, I. E. (1998). Assignment and sequencing models for the scheduling of process systems. Annals of Operations Research, 81, 433–466.

155. Pisinger, D., & Ropke, S. (2010). Large neighborhood search. In M. Gendreau, & J.-Y. Potvin (Eds.), Handbook of metaheuristics (2nd ed., Vol. 10, pp. 399–420). Springer.

156. Pochet, Y., & Wolsey, L. A. (2006). Production planning by mixed integer programming. New York; Berlin: Springer.

157. Pochet, Y., & Warichet, F.(2008). A tighter continuous time formulation for the cyclic scheduling of a mixed plant. Computers & Chemical Engineering, 32, 2723–2744.

158. Potts, C. N. (1980). Analysis of a heuristic for one machine sequencing with release dates and delivery times. Operations Research, 28, 1436–1441.

159. Pranzo, M., Meloni, C., & Pacciarelli, D. (2003). A new class of greedy heuristics for job shop scheduling problems. In K. Jansen, K. Jansen, et al. (Eds.), WEA 2003 (pp. 223–236). Springer LNCS 2647.

160. Prasad, P., & Maravelias, C. T. (2008). Batch selection, assignment and sequencing in multi-stage multi-product processes. Computers & Chemical Engineering, 32, 1106–1119.

161. Prata, A., Oldenburg, J., Marquardt, W., Nystrom, R., & Kroll, A. (2007). Integrated scheduling and dynamic optimization of grade transitions for a continuous polymerization reactor. Computers and Chemical Engineering, 32, 463–476.

162. Relvas, S., Matos, H. A., Barbosa-Povoa, A. P. F. D., & Fialho, J. (2007). Reactive scheduling framework for a multiproduct pipeline with inventory management. Industrial & Engineering Chemistry Research, 46(17), 5659–5672.

163. Régin, J.-C. (2011). Global constraints: A survey. In P. van Hentenryck, & M. Milano (Eds.), Hybrid optimization: The ten years of CPAIOR (pp. 63–134). Springer.

164. Ribas, I., Leisten, R., & Framinan, ˜ J. M.(2010). Review and classification of hybrid flow shop scheduling problems from a production system and a solutions procedure perspective. Computers and Operations Research, 37(8), 1439–1454.

165. Rodosek, R., Wallace, M., & Hajian, M. (1999). A new approach to integrating mixed integer programming and constraint logic programming. Annals of Operations Research, 86, 63–87.

166. Rodrigues, M. T. M., Gimeno, L., Passos, C. A. S., & Campos, M. D. (1996). Reactive scheduling approach for multipurpose chemical batch plants. Computers & Chemical Engineering, 20, S1215–S1220.

167. Roe, B., Papageorgiou, L. G., & Shah, N. (2005). A hybrid MILP/ CLP algorithm for multipurpose batch process scheduling. Computers & Chemical Engineering, 29, 1277–1291.

168. Rossi, F., van Beek, P., & Walsh, T.(Eds.).(2006). Handbook of constraint programming. Amsterdam, The Netherlands: Elsevier.

169. Ryu, J.-H., Dua, V., & Pistikopoulos, E. N. (2007). Proactive scheduling under uncertainty: A parametric optimization approach. Industrial & Engineering Chemistry Research, 46, 8044–8049.

170. Sabuncuoglu, I., & Bayiz, M. (1999). Job shop scheduling with beam search. European Journal on Operational Research, 118, 390–412.

171. Sahinidis, N. (1996). BARON: A general purpose global optimization software package. Journal of Global Optimization, 8(2), 201–205.

172. Sahinidis, N. (2004). Optimization under uncertainty: State-of-the-art and opportunities. Computers & Chemical Engineering, 28, 971–983.

173. Sahinidis, N. V., & Grossmann, I. E. (1991a). MINLP model for cyclic multiproduct scheduling on continuous parallel lines. Computers & Chemical Engineering, 15, 85.

174. Sahinidis, N. V., & Grossmann, I. E.(1991b). Reformulation of multiperiod MILP models for planning and scheduling of chemical processes. Computers & Chemical Engineering, 15(4), 255–272.

175. Samsatli, N. J., & Shah, N. (1996). An optimization based design procedure for biochemical processes.2. Detailed scheduling. Food and Bioproducts Processing, 74(C4), 232–242.

176. Sand, G., & Engell, S. (2004). Modeling, solving real-time scheduling problems by stochastic integer programming. Computers & Chemical Engineering, 28(6–7), 1087–1103.

177. Sand, G., Till, J., Tometzki, T., Urselmann, M., Engell, S., & Emmerich, M. (2008). Engineered versus standard evolutionary

algorithms: A case study in batch scheduling with recourse. Computers and Chemical Engineering, 32, 2706–2722.

178. Säynevirta, S., & Luotojärvi, M.(2004). Integrated paper production and energy planning. In Proceedings, PulPaper 2004 conferences – Energy and carbon management.

179. Schilling, G., & Pantelides, C. C.(1996). A simple continuous-time process scheduling formulation and a novel solution algorithm. Computers & Chemical Engineering, 20, S1221–S1226.

180. Schulte, C., & Tack, G. (2010). Implementing efficient propagation control. In Third workshop on techniques for implementing constraint programming systems (TRICS 2010).

181. Schweiger, C. A., & Floudas, C. A. (1998). Process synthesis, design and control: A mixed integer optimal control framework. In Proceedings of DYCOPS-5 on dynamics and control of process systems (pp. 189–194).

182. Seid, R., & Majozi, T. (2012). A robust mathematical formulation for multipurpose batch plants. Chemical Engineering Science, 68(1), 36–53.

183. Shah, N., Pantelides, C. C., & Sargent, R. (1993). Optimal periodic scheduling of multipurpose batch plants. Annals of Operations Research, 42, 193.

184. Shaik, M., & Floudas, C. A. (2008). Unit-specific event-based continuous-time approach for short-term scheduling of batch plants using RTN framework. Computers & Chemical Engineering, 32, 260–274.

185. Shaik, M., & Floudas, C. (2009). Novel unified modeling approach for short-term scheduling. Industrial & Engineering Chemistry Research, 48, 2947.

186. Shaik, M. A., Floudas, C. A., Kallrath, J., & Pitz, H. (2009). Production scheduling of a large-scale industrial continuous plant: Short-term and medium-term scheduling. Computers & Chemical Engineering, 33, 670–686.

187. Shaw, P. (1998). Using constraint programming and local search methods to solve vehicle routing problems. In Principles and practice of constraint programming (CP 1998), LNCS 1520 (pp. 417–431). Springer.

188. Shaw, P. (2011). Constraint programming and local search hybrids. In P. van Hentenryck, & M. Milano (Eds.), Hybrid optimization: The ten years of CPAIOR (pp. 271–304). Springer.

189. Shobrys, D. E., & White, D. C. (2002). Planning, scheduling and control systems: Why cannot they work together. Computers & Chemical Engineering, 26, 149–160.

190. Simonis, H. (1996). Application development with the CHIP system. In G. M. Kuper, & M. Wallace (Eds.), Constraint databases and applications, ESPRIT WG CONTESSA workshop proceedings, LNCS 1034 (pp. 1–21). Springer.

191. Simonis, H. (2001). Building industrial applications with constraint programming. In H. Comon, C. Marché, & R. Rreinen (Eds.), Constraints in computational logics, Lecture notes in computer science 2002 (pp. 271–309). Springer.

192. Stadtler, H. (2005). Supply chain management and advanced planning – Basics, overview and challenges. European Journal of Operational Research, 163(3), 575–588.

193. Stafford, E. F., Tseng, F. T., & Gupta, J. N. D. (2005). Comparative evaluation of MILP flowshop models. Journal of the Operational Research Society, 56(1), 88–101.

194. Stuckey, P. J., de la Banda, M. G., Maher, M., Marriott, K., Slaney, J., Somogyi, Z., et al. (2005). The G12 project: Mapping solver independent models to efficient solutions. In P. van Beek (Ed.), Principles and practice of constraint programming (CP 2005), LNCS 3668 (pp. 314–327). Springer.

195. Subbiah, S., Schoppmeyer, C., & Engell, S. (2011). An intuitive and efficient approach to process scheduling with sequence-dependent changeovers using timed automata models. Industrial and Engineering Chemistry Research, 50(9), 5131–5152.

196. Subbiah, S., Tometzki, T., Engell, S., & Panek, S. (2009). Multi-product batch plants scheduling with intermediate due dates using priced timed automata models. Computers and Chemical Engineering, 33, 1661–1676.

197. Subrahmanyam, S., Kudva, G. K., Bassett, H. H., & Pekny, J. F. (1996). Application of distributed computing to batch plant design and scheduling. AIChE Journal, 42, 1648–1661.

198. Subramanian, K., Maravelias, C. T., & Rawlings, J. B. (2012). A state-space model for chemical production scheduling. Computers & Chemical Engineering, 47, 97–110.

199. Sundaramoorthy, A., & Karimi, I. A. (2005). A simpler better slot-based continuoustime formulation for short-term scheduling in multipurpose batch plants. Chemical Engineering Science, 60, 2679–2702.

200. Sundaramoorthy, A., & Maravelias, C. T. (2008a). Simultaneous batching and scheduling in multistage multiproduct processes. Industrial & Engineering Chemistry Research, 47, 1546–1555.

201. Sundaramoorthy, A., & Maravelias, C. T.(2008b). Modeling of storage in batching and scheduling of multistage processes. Industrial & Engineering Chemistry Research, 47(17), 6648–6660.

202. Sundaramoorthy, A., & Maravelias, C. T. (2011a). Computational study of network-based mixed-integer programming approaches for chemical production scheduling. Industrial & Engineering Chemistry Research, 50, 5023–5040.

203. Sundaramoorthy, A., & Maravelias, C. T. (2011b). A general framework for process scheduling. AIChE Journal, 57(3), 695–710.

204. Sundaramoorthy, A., Maravelias, C. T., & Prasad, P. (2009). Scheduling of multistage batch processes under utility constraints. Industrial & Engineering Chemistry Research, 48(13), 6050–6058.

205. Susarla, N., Li, J., & Karimi, I. A. (2010). A novel approach to scheduling multipurpose batch plants using unit-slots. AIChE Journal, 56(7), 1859–1879.

206. Terrazas-Moreno, S., Flores-Tlacuahuac, A., & Grossmann, I. E. (2007). Simultaneous cyclic scheduling and optimal control of polymerization reactors. AIChE Journal, 53, 2301–2315.

207. Tili, J., Sand, G., Urselmann, M., & Engell, S. (2007). A hybrid evolutionary algorithm for solving two-stage stochastic integer programs in chemical batch scheduling. Computers & Chemical Engineering, 31, 630–647.

208. Tometzki, T., & Engell, S. (2009). Hybrid evolutionary optimization of two-stage stochastic integer programming problems: An empirical investigation. Evolutionary Computation, 17, 511–526.

209. Tometzki, T., & Engell, S. (2011). Risk conscious solution of planning problems under uncertainty by hybrid multi-objective evolutionary algorithms. Computers & Chemical Engineering, 35(11), 2521–2539.

210. Tseng, F. T., Stafford, E. F., & Gupta, J. N. D. (2004). An empirical analysis of integer programming formulations for the permutation flowshop. Omega-International Journal of Management Science, 32(4), 285–293.

211. van den Heever, S. A., & Grossmann, I. E. (2003). A strategy for the integration of production planning and reactive scheduling in the optimization of a hydrogen supply network. Computers & Chemical Engineering, 27(12), 1813–1839.

212. van Elzakker, M. A. H., Zondervan, E., Raikar, N. B., Grossmann, I. E., & Bongers, P. M. M. (2012). Scheduling in the FMCG industry: An industrial case study. Industrial & Engineering Chemistry Research, 51(22), 7800–7815.

213. van Hentenryck, P., Lustig, I., Michel, L., & Puget, J. F. (1999). The OPL optimization programming language. Cambridge, MA: MIT Press.

214. Velez, S., Sundaramoorthy, A., & Maravelias, C. T. (2013). Valid inequalities based on demand propagation for chemical production scheduling MIP models. AIChE Journal, 59(3), 872–887.

215. Velez, S., & Maravelias, C. T.(2013a). Multiple and nonuniform time grids in discretetime MIP models for chemical production scheduling. Computers & Chemical Engineering, 53, 70–85.

216. Velez, S., & Maravelias, C. T. (2013b). Reformulations and branching methods for mixed-integer programming chemical production scheduling models. Industrial & Engineering Chemistry Research, 52(10), 3832–3841.

217. Velez, S., & Maravelias, C. T. (2013c). A branch-and-bound algorithm for the solution of chemical production scheduling MIP models using parallel computing. Computers & Chemical Engineering, 55, 28–39.

218. Velez, S., & Maravelias, C. T. (2013d). Mixed-integer programming model and tightening methods for scheduling in general chemical production environments. Industrial & Engineering Chemistry Research, 52(9), 3407–3423.

219. Viswanathan, & Grossmann, I. E. (1990). A combined penalty function and outer-approximation method for MINLP optimization. Computers & Chemical Engineering, 14(July (7)), 769–782.

220. Wallace, M., Novello, M. S., & Schimpf, J. (1997). ECLiPSe: A platform for constraint logic programming. ICL Systems Journal, 12, 159–200.

221. Wassick, J.(2009). Enterprise-wide optimization in an integrated chemical complex. Computers & Chemical Engineering, 33, 1950–1963.

222. Wassick, J. M., & Ferrio, J. (2011). Extending the resource task network for industrial applications. Computers & Chemical Engineering, 35(10), 2124–2140.

223. Westerlund, T., & Pettersson, F. (1995). An extended cutting plane method for solving convex MINLP problems. Computers & Chemical Engineering, 19(Suppl.), S131–S136.

224. Westerlund, T., & Pörn,R.(2002). Solving pseudo-convexmixed integer optimization problems by cutting plane techniques. Optimization and Engineering, 3, 253–280.

225. Wolsey, L. A. (1998). Integer programming. New York: Wiley.

226. Wu, D., & Ierapetritou, M. G.(2003). Decomposition approaches for the efficient solution of short-term scheduling problems. Computers & Chemical Engineering, 27, 1261–1276.

227. Yee, K. L., & Shah, N. (1997). Scheduling of multistage fast-moving consumer goods plants. Journal of the Operational Research Society, 48(12), 1201–1214.

228. Yee, K. L., & Shah, N. (1998). Improving the efficiency of discrete time scheduling formulation. Computers & Chemical Engineering, 22(S1), S403–S410.

229. You, F., Castro, P. M., & Grossmann, I. E.(2009). Dinkelbach's algorithm as an efficient method for solving a class of MINLP models for large-scale cyclic scheduling problems. Computers & Chemical Engineering, 33, 1879–1889.

230. Yunes, T., Aron, I., & Hooker, J. N. (2010). An integrated solver for optimization problems. Operations Research, 58, 342–356.

231. Zeballos, L. J., Novas, J. M., & Henning, G. P. (2011). A CP formulation for scheduling multiproductmultistage batch plants. Computers & Chemical Engineering, 35(12), 2973–2989.

232. ZVEI. (2011). Manufacturing Execution Systems (MES): Industry specific requirements and solutions. , ISBN 978-3-939265-23-8 www.zvei.org/automation/mes

Citations

CHAPTER 1

Nazatul Aini Abd Majid, Mark P. Taylor, John J. J. Chen, and Brent R. Young, "Aluminium Process Fault Detection and Diagnosis," Advances in Materials Science and Engineering, vol. 2015, Article ID 682786, 11 pages, 2015. doi:10.1155/2015/682786.

CHAPTER 2

John P. T. Mo, "Performance Assessment of Product Service System from System Architecture Perspectives," Advances in Decision Sciences, vol. 2012, Article ID 640601, 19 pages, 2012. doi:10.1155/2012/640601.

CHAPTER 3

Z. Huang, F. He, A. Zheng, K. Zhao, S. Chang, X. Li, H. Li and Z. Zhao, "Thermodynamic Analysis and Synthesis Gas Generation by Chemical-Looping Gasification of Biomass with Nature Hematite as Oxygen Carriers," Journal of Sustainable Bioenergy Systems, Vol. 3 No. 1, 2013, pp. 33-39. doi: 10.4236/jsbs.2013.31004.

CHAPTER 4

Ghania Henini, Fatiha Souahi, and Ykhlef Laidani, "Methodology of Supervision by Analysis of Thermal Flux for Thermal Conduction of a Batch Chemical Reactor Equipped with a Monofluid Heating/Cooling System," Modelling and Simulation in Engineering, vol. 2012, Article ID 764614, 10 pages, 2012. doi:10.1155/2012/764614.

CHAPTER 5

A. Gadelha, S. Neto, R. Swarnakar and A. Lima, "Thermo-Hydrodynamics of Core-Annular Flow of Water, Heavy Oil and Air Using CFX," Advances in Chemical Engineering and Science, Vol. 3 No. 4A, 2013, pp. 37-45. doi: 10.4236/aces.2013.34A1006.

CHAPTER 6

Shahzad, K. , Saleem, M. , Ghauri, M. , Khan, W. and Akhtar, N. (2014) Modeling and Analysis of SO2 Emissions under Fast Fluidized Bed Conditions Using One Dimensional Model. Advances in Chemical Engineering and Science,4, 327-338. doi: 10.4236/aces.2014.43036.

CHAPTER 7

A. Eid, O. Benlounes, H. Hilal, C. Rabia and S. Hocine, "Dehydrocyclization of n-Hexane over Heteropolyoxometalates Catalysts," Advances in Chemical Engineering and Science, Vol. 3 No. 1, 2013, pp. 82-92. doi: 10.4236/aces.2013.31010.

CHAPTER 8

Manfred Nagl, Bernhard Westfechtel, Ralph Schneider, Tool support for the management of design processes in chemical engineering, Computers & Chemical Engineering, Volume 27, Issue 2, 15 February 2003, Pages 175-197, ISSN 0098-1354, http://dx.doi.org/10.1016/S0098-1354(02)00164-3.

CHAPTER 9

Iiro Harjunkoski, Christos T. Maravelias, Peter Bongers, Pedro M. Castro, Sebastian Engell, Ignacio E. Grossmann, John Hooker, Carlos Méndez, Guido Sand, John Wassick, Scope for industrial applications of production scheduling models and solution methods, Computers & Chemical Engineering, Volume 62, 5 March 2014, Pages 161-193, ISSN 0098-1354, http://dx.doi.org/10.1016/j.compchemeng.2013.12.001.

Index